SpringerBriefs in Earth System Sciences

Series Editors

Gerrit Lohmann
Jorge Rabassa
Justus Notholt
Lawrence A. Mysak
Vikram Unnithan

For further volumes:
http://www.springer.com/series/10032

Kamal Puri · René Redler
Reinhard Budich

Earth System Modelling – Volume 1

Recent Developments and Projects

 Springer

Kamal Puri
The Centre for Australian Weather
and Climate Research, A partnership
between CSIRO and the Bureau
of Meteorology
Melbourne, VIC
Australia

René Redler
Reinhard Budich
Max-Planck-Institut für Meteorologie
Hamburg
Germany

ISSN 2191-589X   ISSN 2191-5903 (electronic)
ISBN 978-3-642-36596-6   ISBN 978-3-642-36597-3 (eBook)
DOI 10.1007/978-3-642-36597-3
Springer Heidelberg New York Dordrecht London

Library of Congress Control Number: 2011938123

Printed on acid-free paper

Springer is part of Springer Science+Business Media (www.springer.com)

Preface

Climate modelling in former times mostly covered the physical processes in the Earth's atmosphere. Nowadays, there is a general agreement that not only physical, but also chemical, biological and, in the near future, economical and sociological—the so-called anthropogenic—processes have to be taken into account on the way towards comprehensive Earth system models. Furthermore these models include the oceans, the land surfaces and, so far to a lesser extent, the Earth's mantle. Between all these components feedback processes have to be described and simulated.

Today, a hierarchy of models exists for Earth system modelling. The spectrum reaches from conceptual models—back of the envelope calculations—over box-, process- or column-models, to Earth system models of intermediate complexity and finally to comprehensive global circulation models of high resolution in space and time. Since the underlying mathematical equations in most cases do not have an analytical solution, they have to be solved numerically. This is only possible by applying sophisticated software tools, which increase in complexity from the simple to the more comprehensive models.

With this series of briefs on "Earth System Modelling" at hand we focus on Earth system models of high complexity. These models need to be designed, assembled, executed, evaluated and described, both in the processes they depict as well as in the results that experiments carried out with them produce. These models are conceptually assembled in a hierarchy of submodels, where process models are linked together to form one component of the Earth system (Atmosphere, Ocean, ...), and these components are then coupled together to Earth system models in different levels of completeness. The software packages of the many process models comprise a few to many thousand lines of code, which results in a high complexity of the task to develop, optimise, maintain and apply these assembled packages.

Running these models is an expensive business. Due to their complexity and the requirements with respect to the ratios of resolution versus extent in time and space, most of the models can only be executed on high-performance computers, commonly called supercomputers. Even on today's supercomputers, typical model experiments take months to conclude. This makes it highly attractive to increase the efficiency of the codes. On the other hand, the lifetime of the codes exceedes the typical lifetime of computing systems and architectures roughly by a factor of 3. This means that the codes need to be portable and adaptable to emerging computing technology. While previously the computing power of single processors—and that of clustered computers—was achieved mainly from increasing clock speeds of the CPUs, currently increases are only exploitable when the application programmer can make best use of the increasing parallelism off-core, on-core and in threads per core. This adds complexity to areas like IO performance, communication between cores or load balancing to the assignment at hand.

All these requirements put high demands on the programmers to apply software development techniques to the code, making it readable, flexible, well structured, portable and reusable, but most of all capable in terms of performance. Fortunately, these requirements match very well an observation from many research centres: due to the typical structure of the staff at the research centres, code development often has to be done by scientific experts, who typically are not computing or software development experts. Therefore, the code they deliver needs a certain amount of quality control to assure fulfilment of the requirements mentioned above. This quality assurance has to be carried out by staff with detailed knowledge and experience in scientific software development and have a mixed background from computing and science.

Since such experts are rare, an approach to ensure high code quality is the introduction of common software infrastructures or frameworks. These entities attempt to deal with the problem by providing certain standards in terms of coding and interfaces, data formats and source management structures, that enable the code developers and the experimenters to deal with their Earth system models in a well acquainted, efficient way. The frameworks foster the exchange of codes between research institutions, model inter-comparison projects so valuable for model development, and the necessary flexibility to the scientists when moving from one institution to another, which is commonplace behaviour these days.

With an increasing awareness about the complexity of these various aspects, scientific programming has emerged as a rather new discipline in the field of Earth system modelling. At the same time, new journals are being launched providing platforms to exchange new ideas and concepts in this field. Up to now we are not aware of any text book addressing this field, tailored to the specific problems the researcher is confronted with. To start a first initiative in this direction, we have compiled a series of six volumes, each dedicated to a specific topic the researcher is confronted with when approaching Earth System Modelling:

Volume 1:   Recent Developments and Projects
Volume 2:   Algorithms, Code Infrastructure and Optimisation
Volume 3:   Coupling Software and Strategies
Volume 4:   IO and Postprocessing
Volume 5:   Tools for configuring, building and running models
Volume 6:   ESM Data Archives in the Times of the Grid

This series aims at bridging the gap between IT solutions and Earth system science. The topics covered provide insight into state-of-the-art software solutions and in particular address coupling software and strategies in regional and global models, coupling infrastructure and data management, strategies and tools for pre- and post-processing and techniques to improve the model performance. Thus the series aims to provide an overview of the concepts of Earth system modelling suitable for a wide audience comprising researchers with some knowledge of the field and to non-experts.

Volume 1 at hand familiarises the reader with the general frameworks and different approaches for assembling Earth system models. Volume 2 highlights major aspects of design issues that are related to the software development, its maintenance and performance. Volume 3 describes different technical attempts from the software point of view to solve the coupled problem. Once the coupled model is running, data are produced and postprocessed (Volume 4). The whole process of creating the software, running the model and processing the output is assembled into a workflow (Volume 5). Volume 6 describes coordinated approaches to archive and retrieve data.

Hamburg, December 2012   René Redler
                                                                            Reinhard Budich

Contents

Contributors

V. Balaji Geophysical Fluid Dynamics Laboratory, Princeton University, Princeton, NJ, USA, e-mail: V.Balaji@noaa.gov

Reinhard Budich Max-Planck-Institut für Meteorologie, Hamburg, Germany, e-mail: reinhard.budich@zmaw.de

Sarah Callaghan STFC-Rutherford Appleton Laboratory, Oxfordshire, UK, e-mail: sarah.callaghan@stfc.ac.uk

Anthony Craig National Center for Atmospheric Research, Boulder, CO, USA, e-mail: tcraig@ucar.edu

Cecilia De Luca NOAA Cooperative Institute for Research in Environmental Sciences, Boulder, CO, USA, e-mail: cecelia.deluca@noaa.gov

Eric Guilyardi NCAS, University of Reading, Berkshire, UK, e-mail: E.D.A. Guilyardi@readinc.ac.uk

Robert Jacob Argonne National Laboratory, Chicago, IL, USA, e-mail: jacob@mcs.anl.gov

Sylvie Joussaume CNRS, Institut Pierre Simon Laplace, Paris, France, e-mail: sylvie.joussaume@lsce.ipsl.fr

Timothy Lenton College of Life and Environmental Sciences, University of Exeter, Exeter, UK, e-mail: T.M.Lenton@exeter.ac.uk

Mark Morgan Institut Pierre Simon Laplace, Université Pierre Marie Curie, Paris, France, e-mail: momipsl@ipsl.jussieu.fr

Kamal Puri The Centre for Australian Weather and Climate Research, A partnership between CSIRO and the Bureau of Meteorology, Melbourne, Australia, e-mail: K.Puri@bom.gov.au

Lois Steenman-Clark NCAS, University of Reading, Berkshire, UK

Akimasa Sumi Integrated Research System for Sustainability Science, Transdisciplinary Initiative for Global Sustainability, Atmosphere and Ocean Research Institute, The University of Tokyo, Tokyo, Japan, e-mail: sumi@ccsr.u-tokyo.ac.jp

Sophie Valcke Centre Européen de Recherche et de Formation Avancée en Calcul Scientifique, Toulouse, France, e-mail: Sophie.Valcke@cerfacs.fr

Mariana Vertenstein National Center for Atmospheric Research, Boulder, CO, USA, e-mail: mvertens@ucar.edu

Chapter 1
Recent Developments and Projects

Kamal Puri

Climate variability and change have a major influence on the social and natural environments, and major climate research programmes have been aimed at determining the extent to which climate can be predicted, and to determine the extent of the human influence on climate. A further aim of the programmes is to achieve a deeper and more quantitative understanding of the role of human perturbations to biogeochemical cycles in altering the coupled climate system, and vice versa.

The past decade has seen major improvements in our ability to provide accurate weather forecasts over the 1–10-day timescales. These improvements are a result of a number of factors such as major increase in the observation network that provides high resolution (in time and space) information on key atmospheric variables, development of analysis and assimilation methods that allow effective use of these data, improvements to all components of numerical models including increased resolution and development of comprehensive methods to verify and diagnose model output. Despite the notable increase in forecast skill, there are still deficiencies in our ability to accurately predict high-impact weather systems that can have significant impact on society, the economy and the environment; examples of these systems include heavy precipitation, flooding, tropical cyclone landfall, destructive surface winds etc. Improving the skill of high impact weather forecasts is a major scientific and societal challenge.

As noted in the "The World Climate Research Programme Strategic Framework 2005–2015" (WCRP-123 WMO.TD No.1291),[1] developments in atmospheric science and technology provide the opportunity to address the predictability of the total climate system for the benefit of society and to address the seamless prediction of the climate system from weekly weather to seasonal, decadal and centennial climate variations and anthropogenic climate change.

---

[1] http://wcrp.wmo.int/pdf/COPES_document_FINAL.12_September_05.pdf

K. Puri
The Centre for Australian Weather and Climate Research, A partnership between CSIRO and the Bureau of Meteorology, Melbourne, Australia
e-mail: K.Puri@bom.gov.au

K. Puri et al., *Earth System Modelling – Volume 1*, SpringerBriefs in Earth System Sciences, DOI: 10.1007/978-3-642-36597-3_1, © The Author(s) 2013

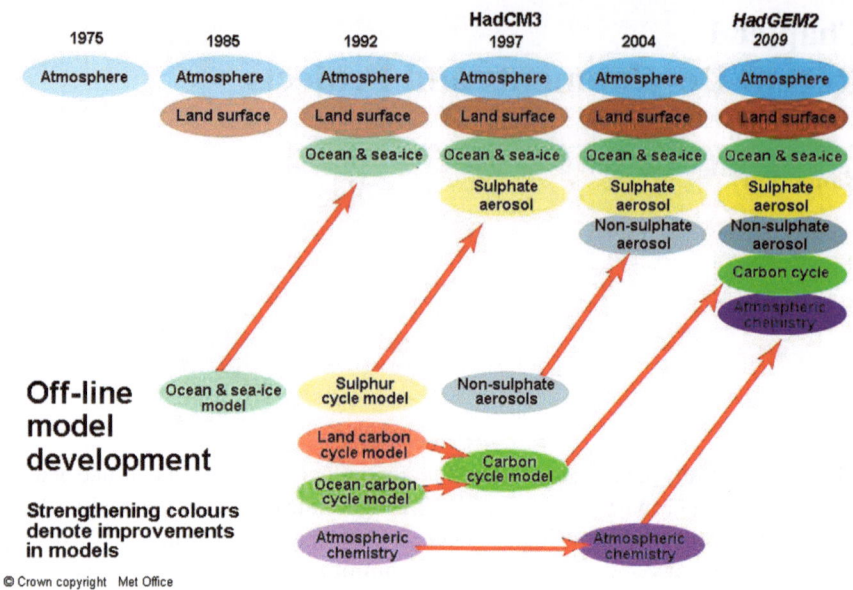

Fig. 1.1 Development of Hadley Centre ESM

Earth system models (ESMs) are essential tools required to address the so-called seamless climate prediction problem. The evolution over time of Earth system models can best be exemplified by the development of the Met Office Hadley Centre for Climate Prediction and Research model as summarised in Fig. 1.1; most major ESMs have undergone very similar developments. A key aspect is the increasing resolution and complexity of the models that now include the atmosphere, land surface, ocean and sea-ice, aerosols, carbon cycle and atmospheric chemistry. Numerical weather prediction (NWP) models have until recently only included the atmosphere and land surface; however in the past few years ocean and sea-ice have been added for seasonal prediction and tropical cyclone prediction. An important aspect of NWP and seasonal prediction systems is the inclusion of complex data analysis and assimilation methods that provide the capability of assimilating data from a wide range of observation systems, and particularly from satellites. Furthermore there is now a growing interest in environmental prediction. Accordingly, some operational centres have already made a move in this direction by implementing an atmospheric chemistry module in the forecast model and extending their data assimilation system to enable assimilation of chemical species (such a system is already being run at the European Centre for Medium Range Weather Forecasts (ECMWF) under the European project Monitoring Atmospheric Composition and Climate, MACC to provide daily analyses and forecasts). Ocean data assimilation is now used routinely and operationally for seasonal prediction. Although coupled data assimilation is currently at an early stage, it is an area that will see major activity in the future.

As can be seen from the above discussion, the scope of the advanced science and technology that underpins leading practice in Earth system modelling is such that it requires significant resources in research and development, in the infrastructure enabling the strategic and tactical application of the science, and in enabling availability of increasing supercomputing power and data storage. Major coordination and focused team research is now necessary to address the breadth and complexity of the scientific and technical issues. This is the clear message from the centres delivering the most successful outcomes. Accordingly substantial efforts have been expended at the major centres running ESMs in developing the complex infrastructure needed to (1) assemble Earth system coupled models using a coupler, (2) launch/monitor complex simulations including ensembles, (3) access, analyse and share results across a wide community, (4) define and promote technical and scientific standards for earth system modelling, and (5) enable efficient running of the models on new computing architectures which comprise clusters with tens of thousands of cores (several hundred thousands in the future).

The level of resources required to develop and maintain ESMs is such that it is becoming increasingly difficult for individual institutions to provide the resources

Table 1.1 Modules of the ACCESS ESM

Module	Name	Source
Atmosphere	UK Met Office Unified Model (UM) (Cullen 1993; Davies et al. 2005)	UK Met Office
Ocean	Modular ocean model version 4 (MOM4) (Griffies 2004; Griffies et al 2004)	NOAA Geophysical Fluid Dynamics Laboratory
Sea-ice	The Los Alamos Sea ice model version 4 (CICE4) (Hunke and Dukowicz 1997)	Los Alamos National Laboratory
Land surface/ Carbon cycle	CSIRO atmosphere biosphere Land exchange model (CABLE) (Kowalczyk et al 2006)	CSIRO
Chemistry and Aerosols	United Kingdom Chemistry and Aerosol model (UKCA) (Abraham et al 2012)	UK Met Office, Leeds and Cambridge Universities
Data assimilation — atmosphere	4-dimensional variational assimilation (4DVAR) (Rawlins et al 2007)	UK Met Office
Data Assimilation — ocean	Ensemble Kalman Filter (Oke et al 2005)	Bureau of Meteorology/CSIRO
Coupler	OASIS	Centre Européen de Recherche et de Formation Avancée en Calcul Scientifique (CERFACS)

required to develop and sustain a state-of-the-art system. One possible means to address this major issue is to pool resources and to source modules comprising the ESM in-house or from external sources. A recent example of such a development is the decision in Australia by the Bureau of Meteorology and the Commonwealth Scientific and Industrial Research Organisation (CSIRO) to jointly develop their ESM. The fully coupled system called the Australian Community Climate and Earth System Simulator (ACCESS; Puri 2012) is aimed at developing a unified system for weather prediction for all time and space scales and climate/climate change simulations. The component modules of ACCESS and their sources are shown in Table 1.1. A feature to note is that the modules include in-house developments such as the land surface and carbon cycle module and imported modules such as the atmosphere and data assimilation modules from the Met Office and the ocean module from GFDL. The atmosphere and ocean/sea-ice modules are coupled using the OASIS coupler developed under the PRISM project.

Development of a complex modelling system such as ACCESS has been greatly facilitated by the ready availability of modelling infrastructures noted above (and modules from the developing centres). The following chapters are devoted towards providing the rationale and status of these modelling and infrastructure developments.

References

Cullen M (1993) The unified forecast/climate model. Meteorol Mag 122:81–94

Davies T, Cullen M, Malcolm A, Mawson M, Staniforth A, White A, Wood N (2005) A new dynamical core for the Met Office's global and regional modelling of the atmosphere. Quaterly J Roy Met Soc 131:1759–1782

Griffies S (2004) Fundamentals of ocean climate models. Princeton University Press, Princeton NJ

Griffies S, Harrison M, Pacanowski R, Rosati A (2004) A technical guide to MOM4. Technical report no 5, NOAA/Geophysical Fluid Dynamics Laboratory, GFDL Ocean Group

Hunke EC, Dukowicz JK (1997) An elastic-viscous-plastic model for sea ice dynamics. J Phys Oceanograph 27:1849–1867

Kowalczyk E, Wang Y, Law R, Davies H, McGregor J, Abramowitz G (2006) The CSIRO atmospheric biosphere land exchange (CABLE) model for use in climate models and as an offline model. CSIRO Marine and Atmos Res Paper 013 CSIRO, Australia

Abraham N, Archibald A, Bellouin N, Boucher O, Braesicke P, Bushell A, Carslow K, Collins B, Dalvi M, Emmerson K, Folberth G, Haywood J, Johnson C, Kipling Z, Macintyre H, Mann G, Telford P, Merikanto J, Morgenstern O, O'Connor F, Ordonez C, Osprey S, Pringle K, Pyle J, Rae J, Reddington C, Savage S, Spracken D, Stier P, Rosalind W (2012) Unified model documentation paper no. 84: United Kingdom Chemistry and Aerosol (UKCA) technical description MetUM version 8.2. Technical report no 84, UK Met. Office

Rawlins F, Ballard S, Bovis K, Clayton A, Li D, Inverarity G, Lorenc A, Payne T (2007) The Met Office global four-dimensional variational data assimilation scheme. Quart J Roy Met Soc 133:347–362

Oke P, Schiller A, Griffin G, Brassington G (2005) Ensemble data assimilation for an eddy-resolving ocean model. Q J Roy Met Soc 131:3301–3311

Puri K (2012) The Australian community climate and earth system simulator. Scientific justification and options for system development. Aust Met Oceanogr J Accepted for publication

Chapter 2
The Infrastructure Project of the European Network for Earth System Modelling: IS-ENES

Sylvie Joussaume and Reinhard Budich

2.1 General Overview

The European Network for Earth System Modelling, ENES, gathers the European community developing and applying climate models of the Earth system. This community aims to better understand present and past observed climates and predict future changes under given boundary conditions of anthropogenic and natural forcing. It is strongly involved in the assessments of the Intergovernmental Panel on Climate Change (IPCC) and provides the predictions on which European Union (EU) mitigation and adaptation policies are based. The EU funded seventh Framework Programme (FP7) project IS-ENES,[1] which is finishing its first phase (2009–2013) and preparing for its second phase (2013–2016), aims to promote the development of a common distributed modelling research infrastructure in Europe in order to facilitate the development and exploitation of climate models and better fulfil the societal needs with regards to climate change issues.

The infrastructure dimension of Earth System modelling encompasses models themselves and their associated software, provision and access to model data, and provision and use of high-performance computing (HPC). IS-ENES focuses on all three issues, but does not include the provision of HPC hardware. IS-ENES is a distributed e-infrastructure with a central resource, the "ENES Portal",[2] which provides access to models, tools and model data archives. The main objectives of the IS-ENES project are:

---

[1] http://is.enes.org
[2] http://www.enes.org

S. Joussaume · R. Budich
CNRS, Institut Pierre Simon Laplace, Paris, France
e-mail: sylvie.joussaume@lsce.ipsl.fr

R. Budich
Max-Planck-Institut für Meteorologie, Bundesstr. 53, 20146 Hamburg, Germany
e-mail: reinhard.budich@zmaw.de

- The integration of the European climate and Earth system modelling community;
- The development of Earth System Models for the understanding of climate change;
- High-end simulations enabling better understanding and prediction of future climate change;
- The application of Earth system model simulations to better predict and understand future climate change impacts.

2.2 Rationale and History

The ENES network was established in 2001 following a first foresight exercise performed by the community in 1998 within the EU Euroclivar concerted action (Komen et al. 1998) which recommended "*a better integration of the European modelling effort with respect to human potential, hardware and software*" and "*to develop collaboration, to establish a European climate computing facility, and to enhance the exchange of software and model results*". These recommendations led to the establishment of ENES, as the European Climate Modelling Group advocated by Euroclivar. ENES set up the FP5 infrastructure project "Program for Integrated Earth System Modelling" (PRISM, see Chap. 4). PRISM carried out a successful first step towards the Euroclivar recommendations, establishing a network of expertise around ESM software environments and promoting a standard technical coupling interface now used world-wide, the OASIS coupler. This effort has helped to build the FP7 IS-ENES project (2009–2013) and the FP7 e-Infrastructure project, METAFOR, "Common Metadata for Climate Modelling Digital repositories"[3] (2008–2011, see Chap. 3). Networking within ENES also led to collaborative projects funded within the EU Environment Program, dealing with future climate scenarios, Earth system model developments and evaluation of cloud processes and feedbacks, such as the FP6 ENSEMBLES[4] and FP7 COMBINE[5] and EMBRACE[6] projects.

2.3 Partnerships and Organization

ENES gathers more than 40 institutions and organizations through a Memorandum of Understanding. These include climate modelling centres, computing centres and manufacturers, data centres and centres with expertise in computational science and technology. ENES is governed through a Scientific Advisory Board. An ENES HPC task force has been set up to elaborate a common strategy on HPC and to help provide a common interface to the European high-performance

---

[3] http://metaforclimate.eu

[4] http://www.ensembles-eu.org

[5] http://www.combine-project.eu

[6] http://www.smhi.se/embrace

computing infrastructure PRACE (Partnership for Advanced Computing in Europe).[7] It is anticipated that a similar group on data aspects will be established in the future, interfacing ENES with the international Earth System Grid (ESG) partnership. Within IS-ENES, a subset of 18 ENES partners from 10 European countries is involved in the project. IS-ENES gathers the European global Earth System Models participating in the Coupled Model Intercomparison Project Phase 5 (CMIP5) namely, the UK Met Office model, the German COSMOS/MPI-ESM model developed at the Max Planck Institute of Hamburg, the French Institut Pierre Simon Laplace (IPSL) and Meteo-France models, the Italian Centro Euro-Mediterraneo per i Cambiamenti Climatici climate model, the EC-Earth model developed from the ECMWF model by a consortium of 8 countries, and, now joining in IS-ENES2, the Norwegian ESM. To address issues related to HPC, several computing centres are involved including the Deutsches Klimarechenzentrum (DKRZ), the Barcelona Supercomputing Centre and the Swedish National Supercomputing Centre, as well as centres dedicated to computational science such as CERFACS and University of Manchester. Work is organized through networking, service and joint research activities around models, HPC and data.

2.4 Overview of Results

The integration of the European Earth System modelling community is an important ambition of IS-ENES. The strategy of the community for the next ten years regarding the European infrastructure needs for ESM was issued in 2012 (ENES 2012). Integration has also been developed through more specific objectives such as the development of the ENES Portal which provide access to IS-ENES services, information relevant for the community, and conducted the first European school on ESMs for young scientists. In order to foster the development of ESMs, IS-ENES has started a support service on modelling including access to the European ocean modelling platform NEMO,[8] the OASIS Coupler and the Climate Data Operators post-processing tools (CDOs). The ENES portal developed within IS-ENES also provides access to climate model documentation and on tools and datasets used for model evaluation. Access to world-class high-performance computing is important for the climate modelling community in order to better understand climate and its dynamics and provide information to society. In order to fulfil this objective, IS-ENES has developed collaborations with PRACE to help the preparation of climate models on PRACE machines. Developments have also been made on improving input/output data exchange and coupling software, which are two bottlenecks for massively parallel simulations. A significant effort has been devoted to the deployment of the European contribution to CMIP5. As a result, European nodes have been among the first to publish data, the first to meet agreed data access standards

---

[7] http://prace-project.eu

[8] http://www.nemo-ocean.eu

and the first to complete the second level of the 3 level quality control process. Through IS-ENES, the two European data portals, the British Atmospheric Data Centre and the DKRZ data centre, have developed quality control of ESG federation data archives. A prototype of a portal for the impacts community has been developed that provides access to documented use cases (see Chap. 3) on methodology aspects.

2.5 Users

The most important users are clearly from the global climate modelling community. However, the developments in the IS-ENES project also serve the regional climate modelling community through sharing of common tools with the global climate modellers and analysts. Thus, for example, these developments have resulted in the coupler OASIS being increasingly used to develop coupled regional Earth System Models and the CDOs. Another important target group is the impacts community that is extensively using the results from climate simulations to force different sectorial impact models. The prototype portal for the impacts community aims at describing how best to use results from climate models and provide visualisation and targeted access to model results. More generally, the ENES infrastructure is expected to improve use and usability of climate models and the dissemination of their results for the benefit of society.

2.6 Future Plans

The 2012 ENES strategy has guided the elaboration of the second phase of IS-ENES. IS-ENES2 is built upon the outputs of both the FP7 IS-ENES and METAFOR projects. It will further integrate the European climate modelling community, stimulate common developments of software for models and their environments, foster the execution and exploitation of high-end simulations, and support the dissemination of model results to the climate research and impacts communities. More specifically, phase 2 will extend ENES services on data from global to regional climate models by supporting results from the World Climate Research Program "Coordinated Regional climate Downscaling Experiments" (CORDEX). The project will also support further metadata developments and ease access to climate projections for the climate impacts community with a common portal for global and regional climate models, including guidance and downscaling tools. Furthermore, common high-resolution modelling experiments for the large European computing facilities will be prepared by the project, underpinning the community's efforts to prepare for the challenge of future exascale computing architectures. The climate modelling community is facing a strong challenge to provide more reliable information to society to prepare for adaptation to climate change. It is expected that this European infrastructure will contribute to it.

References

Komen G, Anderson D, Bengtsson L, Delecluse P, Duplessy JC, Fichefet T, Joussaume S, Jouzel J, Komen G, Latif M, Laursen L, Le Treut H, Mitchell J, Navarra A, Palmer T, Planton S, Ruiz de Elvira A, Schott F, Slingo J, Willebrand J (1998) Climate variability and predictability research in Europe 1999–2004: Euroclivar Recommandations. KNMI, De Bilt, The Netherlands. http://www.knmi.nl/euroclivar/frsum.html

ENES (2012) Infrastructure strategy for the European Earth system modelling community 2012–2022. In: Mitchell J, Budich R, Joussaume S, Lawrence BB, Marotzke J (eds) https://is.enes.org/the-project/communication/ENES%20foresight.pdf/view

Chapter 3
The METAFOR Project

Sarah Callaghan, Eric Guilyardi, Lois Steenman-Clark and Mark Morgan

3.1 General Overview

Mention has been made earlier book about the Common Metadata for Climate Modelling Digital Repositories (METAFOR) project, which is an EU Framework 7 project with the aim of developing a Common Information Model (CIM) to describe climate data and the software models and modelling environments that produce this data in a standard way. Climate modelling is a complex process, requiring complete and accurate metadata (data describing data) to accurately characterise and differentiate between different climate model datasets. The CIM extends the metadata used in existing data repositories and addresses issues like metadata fragmentation, gaps and duplication.

METAFOR aims to ensure the widespread adoption of the CIM through the development of CIM tools and services which will allow climate modellers to (for example) discover, view and difference CIM documents, allowing them to compare data from different climate model runs in a scientifically meaningful way. The CIM will optimise the way climate data infrastructures are used to store knowledge, adding value to primary research data for an increasingly wide range of stakeholders, including non-scientists such as local government and the general public.

---

the METAFOR project team.

S. Callaghan
British Atmospheric Data Centre, STFC - Rutherford Appleton Laboratory, Oxfordshire, UK
e-mail: sarah.callaghan@stfc.ac.uk

E. Guilyardi · L. Steenman-Clark
NCAS, University of Reading, Reading, UK
e-mail: E.D.A.Guilyardi@readinc.ac.uk

M. Morgan
Institut Pierre Simon Laplace (IPSL), Universite Pierre Marie Curie, Paris, France
e-mail: momipsl@ipsl.jussieu.fr

K. Puri et al., *Earth System Modelling – Volume 1*, SpringerBriefs in Earth System Sciences, DOI: 10.1007/978-3-642-36597-3_3, © The Author(s) 2013

3.2 Rationale and History

Climate science plays an increasingly important role for policy-makers, who are faced with the problem of strategic planning at many levels to address the impacts of climate change. In order to support both basic research and effective strategies to mitigate climate change and deal with its impact on society, a wide range of experts from multiple disciplines need both access to data and advice on the suitability of that data for their purposes. This requires better communication of information about available climate resources, particularly data from model projections for the next decades and centuries. This information about climate models is an essential resource for understanding and evaluating climate data, but unfortunately at this time it is not readily available to end users.

The data from climate model runs are currently stored in climate repositories which are poorly connected to each other, with the result that scientists and policy makers are often not aware of what data are available or from what sources. Even if users are aware that the data they need exist and know how to find them, they then often have to deal with a variety of institution-dependent data information models (i.e. file formats, metadata structures, documentation methodologies, etc.). As a consequence, comparing and contrasting just the information about the data, let alone the data itself, is difficult without significant specific expertise.

Climate models generate large amounts of data, hence exploiting that data is also a significant scientific and technical challenge. The models themselves are complex, with each climate model run potentially involving several component models (e.g. atmosphere, ocean, sea-ice, vegetation, land ice, ocean biogeochemistry, atmosphere chemistry) coupled together. Any one of those component models can be configured in many different ways, including not only different parameter values but also changes to the source code itself. Component models, or even compositions of component models, can have multiple versions, and individual component models can be coupled together and run in a multitude of different ways. This range of variability is immense and largely under-documented in the output data, yet this information should be collected and stored in order to ensure the scientific replicability of, and provide confidence in the model runs.

Different model or data users may want to focus on different aspects of the modelling process, depending on their needs and interests. However, up until now, there has been no standard way of describing climate models and the way they are configured, coupled together, and run. This type of information is essential for making accurate comparisons across datasets, and to prevent misinterpretation or misuse of data. The primary aim of METAFOR was to provide a Common Information Model (CIM) to document climate models, their configuration, coupling and software, thus helping to increase confidence in climate model data and the use that policy makers, planners, scientists or industry make of that data.

3.3 Partnerships and Organization

The METAFOR project is a Europe-US collaboration funded by the EU 7th Framework Programme as an e-infrastructure (project #211753). This 2.5M€ project, with 12 partner institutions, is led by Prof. Eric Guilyardi from NCAS-Climate/University of Reading and project managed by Dr. Sarah Callaghan from the British Atmospheric Data Centre (BADC).

The METAFOR consortium assembles all the major groups involved in software infrastructure for climate modelling in Europe. The partners have a long history of collaboration as most worked together in the PRISM FP5 project, and then in the core-funded PRISM Support Initiative (PSI, see Chap. 4). These partners include the University of Reading, CERFACS, Institut Pierre-Simon Laplace (IPSL), the German Climate Computing Centre (DKRZ) and the UK Met Office. They are joined by the BADC to strengthen the service component of PRISM, while the University of Manchester provides extensive technical expertise in high abstraction design, analysis and programming applied to computational science and engineering applications. Météo-France provides the link to and requirements from the seasonal forecast, regional scale and numerical weather prediction (NWP) communities and repositories. Testing of the populated CIM database is performed both by CLIMPACT, an enterprise with long time experience in seasonal and climate data use for specific industrial applications, and other end user groups in the scientific community including Administratia Nationala de Meteorologie, Romania (NMA). The extensibility of the CIM is tested by the University of Cantabria through their expertise in climate impact tools, namely via the FP6 ENSEMBLES downscaling portal.

Finally, as the main outcome of the project is a set of international standards (embodied by the CIM), collaboration with US and international groups is vital. These include CICS / Princeton University, USA, PCMDI (Livermore, USA, in charge of the IPCC multi-model databases) and NCAR (Boulder, USA).

3.4 Overview of Results

3.4.1 Common Information Model (CIM)

The main aim of the METAFOR project was the development of a CIM and, as importantly, a conceptual framework to develop the CIM independently of its implementations. There are two parts to the process of developing the CIM. Firstly the conceptual model of the process (the CONCIM) is modelled using UML. Then the CONCIM can be realised as an application schema (APPCIM) which is currently manifested in XSD,[1] though it would be as equally realised in RDF.[2] This split between conceptual and application development has provided an acceptable platform for the CIM development. However it is recognised that the current version of the CIM will not be the final construct as work is in progress to transition to a construction of the CONCIM which will allow more generic serialisation.

The CIM, especially the software package, is intentionally generic. It provides a structure but not the names of the artefacts within this structure. These names need to be determined by the user community and a controlled vocabulary (CV) developed. Each community within the climate modelling domain will need to play a part in this process with CVs being separately governed. The use of CVs in the CIM has been finalised, with structures for the details of a CV for a CIM entity provided so that a validation tool can locate a CV in an external server to check that the CV is being used by a CIM document correctly.

The CIM has been developed and improved following feedback from the climate modelling community. The current version of the CIM is now 1.5 and can be found as open source.[3] Further work on the CIM is ongoing to extend it to other fields, such as downscaling.

The CIM is at the heart of the METAFOR project and has therefore involved all project partners and necessitated extensive collaboration with diverse climate modelling groups using many different climate models both in Europe and the US.

An information model models a process. For METAFOR this covers the whole domain of climate modelling, from the descriptions of the experiments being undertaken, the simulation being run to carry out these experiments, the software models and tools used to implement the simulations and the data required by and generated by the software. The CIM version 1.5 is used in conjunction with the data for the next climate model inter-comparison (CMIP5) for the Intergovernmental Panel on Climate Change (IPCC).

The CMIP5 experiments range from atmosphere-only multi-century simulations to coupled multi-component models undertaking multi-decadal simulations to short hindcast simulations. So the range of experiment and simulation types has been challenging the scope of the METAFOR CIM. This CMIP5 Use Case was therefore a major test of the CIM structure and the capacity of the CIM to accommodate this diversity. The implementation of this Use Case has ensured that CIM version 1.5 is flexible while also being consistent and robust.

---

[1] W3C's XML Schema language, also known as XSD (XML Schema Definition).

[2] Resource Description Framework (RDF) is a family of World Wide Web Consortium (W3C) specifications originally designed as a metadata data model.

[3] http://metaforclimate.eu/trac/browser/CIM

3.4.2 CMIP5 Questionnaire

The METAFOR project was charged by the WGCM[4] via the CMIP panel to define and collect model and experiment metadata for CMIP5 to be used for the next IPCC assessment (due in 2013). To do this the METAFOR team designed, developed and deployed a web-based questionnaire to collect information from the CMIP5 modelling groups on the details of the models used, how the simulations were carried out, how the models conformed to the CMIP5 experiment requirements and what hardware was used to perform the simulations. The aim of the questionnaire is to document the CMIP5 climate models in sufficient detail so that the model run data can be compared in a scientifically meaningful way. These climate model data will be stored in the CMIP5 archives hosted at PCMDI, BADC and MPI-M, in archives that are anticipated to be greater than 1 PB in size, so good metadata is crucial to their effective use and operation.

The CMIP5 metadata questionnaire[5] was launched in Nov 2010, and is now in use by several of the CMIP5 modelling centres.

The operational deployment of the CMIP5 metadata questionnaire occurred after intensive beta testing and interactions with climate scientists in charge of running the CMIP5 simulations, including the set up of a comprehensive user support mechanism. The CMIP5 questionnaire provides an excellent test bed for the CIM tools and services development, as it produces CIM instances, which can then be used to populate a CIM archive. The tools written to display, query, difference and search the CIM archive are now organised into pilot back end and front end services (e.g. portals). This will form an important and a valuable user community resource for climate scientists interested in the CMIP5 model data.

Help systems and documentation have been developed to support the users of the questionnaire. A dedicated CMIP5 Questionnaire helpdesk service has been set up and is contactable by emailing cmip5qhelp@stfc.ac.uk .

The questionnaire support team can communicate with each other outside of the scope of the helpdesk using the email alias cmip5qteam@badc.nerc.ac.uk. A CMIP5 Questionnaire mailing list has also been set up[6] which users join when they register as questionnaire users. This mailing list gives all of the CMIP5 questionnaire users a facility to communicate with one another and to share knowledge. The support team monitors traffic on this mailing list and creates helpdesk queries where appropriate. It is also used to broadcast messages to users of the questionnaire about training sessions, questionnaire downtime etc. The support team also run live online demonstrations of the questionnaire on an arranged basis, and modelling teams are encouraged to participate in these at a convenient time for them.

---

[4] WCRP/CLIVAR Working Group on Coupled Modelling—http://www.clivar.org/organization/wgcm/wgcm.php

[5] http://q.cmip5.ceda.ac.uk

[6] cmip5q@badc.nerc.ac.uk

3.4.3 CIM Web Portal

The CIM web portal is a web application designed to promote usage of the CIM by a diverse user base. The portal assumes a level of familiarity with climate model metadata running from expert to non-existent. It delivers content falling into several functional categories: information, ontology, search, tools and ingest.

Ingest and search functions allow the portal to act as a conduit for the flow of climate related meta-data. By registering CIM compliant metadata servers with the portal a meta-data administrator can make metadata available to the community via the portal's search functions. An integrated CIM viewer tool provides an intuitive means to navigate through metadata returned by searches.

The ontology and tools functions allow interested parties (e.g. software developers) to access technical information regarding the CIM application schema and to perform actions such as validation against instance of CIM compliant documents. These functions therefore are aimed at speeding up the process of adoption of the CIM as a standard.

Information functions are aimed at the broad public. Besides high-level CIM documentation, links are provided to CIM related publications, and use cases are outlined indicating how different members of the community may leverage the CIM.

Technically the CIM web portal has been developed as a rich internet application with a simple and intuitive user interface. It has been built upon solid software development principles such as a layered architecture, separation of concerns, etc. The portal became fully alive in early autumn 2011.

3.4.4 CIM Web Services

CIM web services serve the CIM user community in a somewhat less direct fashion than the CIM web portal. The web services allow institutes to integrate CIM functions and features into their own operational contexts. One such context may be an institutional web portal that wishes to integrate CIM search into their portal. Another such context may be a command line tool that validates against the CIM validation web service.

Essentially functions and features that are found within the CIM web portal are exposed as web service operations. Such operations can be invoked by an array of consumers as outlined in the above examples, indeed the CIM web portal is itself a consumer of CIM web services. The software design paradigm underlying such an approach whereby functional units are exposed as web services is service orientated architecture (SOA)—a key tenet of modern software systems.

The available CIM web services are repository, search and validation. The repository service allows a consumer to upload, update, retrieve and delete CIM instances. The search service supports the retrieval of CIM instances by various facets. The validation service returns validation reports run against uploaded CIM instances. Collectively the web services allow a CIM eco-system to evolve external to the CIM portal.

3.4.5 CIM Tools

A set of tools to manipulate and work with the CIM has also been developed. Such tools are typically invoked from the command line although they themselves often consume functionality provided by the CIM web services. The tools at present support validation, search and ingestion. The validation and search tools simply hook into the corresponding web services. The ingestion tools parse metadata servers, e.g. a CMIP5 questionnaire atom feed or a Thematic Realtime Environmental Distributed Data Services (THREDDS) catalog, and upload ingested instances to the CIM repository via the CIM repository web service.

The CIM has also been developed so as to be usable as input information for other tools used in the climate community, such as the OASIS coupler or the Bespoke Framework Generator (BFG, see Ford and Riley 2012). The former tool is a climate model coupler that can use the CIM as configuration information for the coupling exchanges it manages (source, target, coupling frequency, transformation, etc.). The latter tool generates framework code to wrap around existing climate models by using the CIM to describe those models, how they couple, and how they should be deployed.

3.5 Users

The Common Information Model (CIM) is primarily aimed at climate modellers, as these are the users who are most likely to take advantage of the CIM to document the results of their model runs. However, tools built using the CIM structure to discover and interrogate CIM instances will allow a far wider range of user to access the climate model metadata and data. These users would include local and national governments and policy makers, and academics working in the impacts and adaptation areas of climate change science.

A wide range of commercial organisations are also becoming interested in climate change issues. Increasingly, these private sector companies need access to primary climate model data to inform decision makers in their own domain or that of their clients. The improved access to the climate data repositories hence represents a clear economic opportunity for Europe. This requires that the specific needs of these key stakeholders be taken into account when exposing climate data resources to a wide audience. The METAFOR project has done this by taking into account that users of the CIM portals and services may range from climate modelling experts to members of the general public.

3.6 Future Plans

Even though the final CIM (for the METAFOR project) has been delivered and is being used within international modelling projects such as CMIP5, it is essential to realise that further work is required. The CONCIM will be brought more in line with existing ISO standards to allow the APPCIM to be realised through existing serialisation approaches rather than the current independent tool that uses XSL transformations into an XSD. Other work is planned to continue to support the use of the CIM in CMIP5 as the CMIP5 exercise continues and to support CMIP5 scientists producing CIM instances and the exploitation of these CIM instances in different portals.

Ensuring the uptake and longevity of the CIM is a key aim of the project; hence work will continue to facilitate the use of the CIM for exploitation by other communities and within other currently funded projects, for example IS-ENES . Governance structures are currently being put in place to ensure the longevity and future development of the CIM and controlled vocabularies, as these are METAFOR project outputs of great value to the scientific community.

Reference

Ford R, Riley G (2012) The bespoke framework generator. In: Ford R, Riley G, Budich R, Redler R (eds) Earth system modelling, recent developments and projects, vol 3. Springer, Heidelberg, pp 55–67

Chapter 4
The European PRISM Network

Sophie Valcke and Eric Guilyardi

4.1 General Overview

The increasing complexity of Earth system models and the computing facilities needed to run them put a heavy technical burden on the research teams active in climate modelling. To ease this burden, the Partnership for Research Infrastructures in Earth System Modelling[1] (PRISM) was initiated by the European Network for Earth System Modelling[2] (ENES) in 2001 as a European Union Framework Programme 5 (EU FP5) project. The PRISM concept was to organize a network of experts in order to share the development, maintenance and support of Earth system modelling software tools and community standards. It was envisioned that the use of specific common standards and tools would reduce the technical development efforts of individual research teams, facilitate the assembling, running, and post-processing of Earth system models, and hence facilitate scientific collaboration between the different research groups in Europe and elsewhere.

4.2 Rationale and History

PRISM was initiated as a European project under the Framework Programme 5 funded for 4.8 MEuros in 2001–2004. The PRISM concept, originally a Euroclivar recommendation, was to enhance what Earth system modellers had in common

---

[1] http://prism.enes.org
[2] http://www.enes.org

S. Valcke
CERFACS, Centre Européen de Recherche et de Formatio Avancée en Calcul Scientifique,
Toulouse, France
e-mail: Sophie.Valcke@cerfacs.fr

E. Guilyardi
NCAS, University of Reading, Reading, UK
e-mail: E.D.A.Guilyardi@readinc.ac.uk

K. Puri et al., *Earth System Modelling – Volume 1*, SpringerBriefs in Earth System Sciences, DOI: 10.1007/978-3-642-36597-3_4, © The Author(s) 2013

(compilers, message passing libraries, algebra libraries, etc.) and to share the development, maintenance and support of a wider set of Earth System Modelling (ESM) software tools and standards. The main objective behind the concept was to reduce the technical development efforts of individual research teams, facilitate the assembling, running, monitoring, and post-processing of ESMs based on state-of-the-art component models developed at the different climate research centres in Europe and elsewhere, and thereby promote the scientific diversity of the climate modelling community. As demonstrated within the Ocean Atmosphere Sea Ice Soil (OASIS) user community (see below), sharing software tools also provides a powerful incentive for increased scientific collaboration. Furthermore it stimulates computer manufacturers to contribute towards increasing the tool portability and the optimisation of next generation of platforms for ESM needs.

The extensive use of the OASIS coupler illustrates the benefits of a successful shared software infrastructure. In 1991, the Centre Européen de Recherche et Formation Avancée en Calcul Scientifique (CERFACS) decided to develop a software for coupling different geophysical component models developed independently by different research groups. The OASIS development team strongly focussed on efficient user support and on the constant integration of the developments fed back by the users. This strong interaction has resulted in a constantly growing community. OASIS now capitalises on about 40 person-years (py) of mutual developments and fulfils the coupling needs of about 35 climate research groups around the world. The effort invested therefore represents, on a first order, 40 py/35 groups = 1,1 py/group, which is certainly much less than the effort that would have been required by each group to develop its own coupler. PRISM represented the first major collective effort, at the European level, to develop ESM supporting software in a shared and coherent way. This effort has been recognised by the Joint Scientific Committee (JSC) and the Modelling Panel of the World Climate Research Programme (WCRP) that has endorsed it as a "key European infrastructure project".

4.3 Partnerships and Organization

Over the 2005-2009 period, support and maintenance of PRISM products was ensured through multi-institute funding from 16 key European institutions and computer manufacturers contributing to the so-called PRISM Support Initiative. The seven main partners of the PRISM Support Initiative were : CERFACS (France), NEC Laboratories Europe - IT Research Division (NLE-IT, Germany), Centre National de la Recherche Scientifique (CNRS, France), Gruppe Modelle & Daten from the Max-Planck-Institut für Meteorologie (MPI-M M&D, Germany), European Centre for Medium-Range Weather Forecasts (ECMWF), National Centre for Atmospheric Science from the University of Reading (NCAS, UK) and the Met Office (UK).

PRISM was overseen by the PRISM Steering Board (one member per partner) that reviewed work plans proposed by the PRISM Core Group composed of PRISM coordinator(s), the leaders of the PRISM Areas of Expertise (see sect. 4.4), and the chair of the PRISM User Group. The PRISM User Group was composed of all climate modelling groups using the PRISM software tools.

The PRISM Support Initiative successfully set up two European funded projects, METAFOR (see Chap. 3) and IS-ENES (see Chap. 2), in which most of the PRISM activity is occurring today.

4.4 Overview of Results

The PRISM Support Initiative was organised around five areas of expertise that addressed metadata, code coupling, integration and modelling environments, data processing and management, and computing issues.

Each of the five PRISM areas of expertise (PAEs) had the following remits:

- Promote and, if needed, develop software tools for Earth System Modelling. A PRISM tool must be portable, usable independently and interoperable with the other PRISM tools, and freely available for research. There should be documented interest from the community to use the tool and the tool developers must be ready to provide user support;
- Encourage and organise a related network of experts, including technology watch;
- Promote and participate in the definition of community standards where needed;
- Coordinate with other PRISM areas of expertise and related international activities.

The scope of the PAE "Code coupling and I/O" was to develop, maintain, and support the OASIS couplers and also to keep strong relations with groups developing different technical approaches for code coupling, for example PALM (Buis et al. 2006), Earth System Modeling Framework (ESMF) (see Chap. 5), and BFG (see Chap. 7 in Volume 3 of this series). In IS-ENES, these tasks are currently ensured within workpackage "Earth System Models, Tools and Environments : Development and Integration".

The PAE "Integration and modelling environments" targeted source version control for software and model development, code extraction and compilation, (coupled) model run configuration, job running, and integration with archive systems. This PAE therefore had interest in tools like the UK Met Office Flexible Configuration Management Software (FCM) also used at IPSL, the Standard Compiling and Running Environments developed at MPI-M&D (SCE and SRE), and the ECMWF job management and flow control tool Supervisor Monitor Scheduler (SMS).

The overall objective of the PAE "Data processing, visualisation and management" was the development of standards and infrastructure for data processing, (possibly geographically distributed) archiving and exchange in Earth system research. For data processing, the main interest was in the Climate Data Operator (CDO) processing tool developed at MPI-M. Distributed databases and archiving is the main focus of IS-ENES workpackages 5 and 10.

Metadata has in the last few years become a key component of new schemas and ideas to promote the interchangeability of Earth system models or modelling components as well as data. The "Metadata" PAE provided a forum to discuss, coordinate or interact with national and international efforts addressing metadata issues, such as the NCAS-Climate Numerical Model Metadata (NMM), the Earth System Curator in the US, or the Network Common Data Form (NetCDF) Climate and Forecast (CF) convention. The activity of this PAE has now been integrated in the METAFOR project.

While computer vendors have to be kept informed about requirements emerging from the climate modelling community, the climate modellers need to be kept informed about the evolution of computing platforms. PRISM played an active role in this by forming a new PAE "Computing" devoted to supercomputing technology trends. Technical topics include file IO, algorithmic development, and portable software for parallel and vector systems. These activities are currently driven by the IS-ENES "HPC Task Force".

4.5 Users

The strength and success of PRISM was to set-up a network allowing Earth system model developers to share their technical expertise and experiences. The different partners (see Sect. 4.3) and the whole ENES community naturally participated in setting up this network, which is currently being strengthened further under the IS-ENES umbrella.

As detailed above PRISM also promoted the use of common standard tools, in particular the OASIS coupler used by approximately 35 climate modelling groups around the world, and the CDO processing tool having users world-wide. This corresponds directly to one of the main tasks of IS-ENES workpackage "Strengthening the European Network on Earth System Modelling".

4.6 Conclusions

The benefit of the PRISM Support Initiative was to allow "best of breed" software tools to emerge naturally. In the areas where unique standards have to be pre-defined, for example for metadata definition, the key role of PRISM was to provide the European entry point for international coordination within the World Climate Research Programme (WCRP).

The PRISM network was naturally well placed to ensure sustainability of the developments made in specific projects and to seek additional funding to support more networking activities and technical developments. The effectiveness of the role played by PRISM in this activity contributed in a significant way to the successful funding of the METAFOR (see Chap. 3) and IS-ENES (see Chap. 2) projects.

The objective of IS-ENES and its recently funded follow-on project IS-ENES2 is to establish a comprehensive e-infrastructure providing an easy-to-use and centralized access to the different resources needed for Earth System modelling. This e-infrastructure and its three main dimensions, namely the European ESMs and their associated software tools, high-performance computing and ESM data archives, now naturally replace PRISM as the driving force towards developing a common infrastructure for climate modelling in Europe.

IS-ENES2 will build on the achievements of PRISM, METAFOR and IS-ENES to strengthen the network of technical expertise on European ESMs and the development and sharing of standards and tools across different European climate modelling organisations.

Reference

Buis S, Piacentini A, Déclat D (2006) PALM: a computational framework for assembling high performance computing applications. Concurrency Computat: Pract Exper 18(2):247–262

Chapter 5
The Earth System Modeling Framework

Cecelia DeLuca and V. Balaji

5.1 General Overview

The Earth System Modeling Framework (ESMF)[1] is open source software for building climate and weather related modeling components, and coupling them together to form applications. ESMF was motivated by the desire to exchange modeling components amongst *modelling* centres and to reduce costs and effort by sharing codes.

The ESMF package is comprised of a *superstructure* of coupling tools and component wrappers with standard interfaces, and an *infrastructure* of utilities for common functions, including calendar management, message logging, grid transformations, and data communications. The project is distinguished by its strong emphasis on community governance and distributed development, and by a diverse customer base that includes modeling groups from universities, major U.S. research centres, the National Weather Service, the Department of Defense, and the National Aeronautics and Space Administration (NASA). The ESMF development team is centered at the National Center for Atmospheric Research (NCAR).

5.2 Rationale and History

The ESMF collaboration had its roots in the Common Modeling Infrastructure Working Group (CMIWG), an unfunded, grass-roots effort to explore ways of enhancing collaborative Earth system model development. The CMIWG attracted

---

[1] http://www.esmf.ucar.edu

C. DeLuca
NOAA Cooperative Institute for Research in Environmental Sciences, Boulder, CO, USA
e-mail: cecelia.deluca@noaa.gov

V. Balaji
Geophysical Fluid Dynamics Laboratory, Princeton University, Princeton, NJ, USA
e-mail: V.Balaji@noaa.gov

K. Puri et al., *Earth System Modelling – Volume 1*, SpringerBriefs in Earth System Sciences, DOI: 10.1007/978-3-642-36597-3_5, © The Author(s) 2013

broad participation from U.S. weather and climate modeling groups at research and operational centers. In a series of meetings held from 1998 to 2000, CMIWG members established general requirements and a preliminary design for a common software framework.

In September 2000, the NASA Earth Science Technology Office (ESTO) released a solicitation that called for the creation of an "Earth System Modeling Framework." A critical mass of CMIWG participants agreed to develop a coordinated response, based on their strawman framework design, and submitted three linked proposals. The first focused on development of the core ESMF software, the second on deployment of Earth science modelling applications, and the third on deployment of ESMF data assimilation applications. All three proposals were funded, at a collective level of 9.8M over a three year period. As the ESMF project gained momentum, it replaced the CMIWG as the focal point for developing community modeling infrastructure in the U.S.

During the period of NASA funding, the ESMF team developed a prototype of the framework and used it in a number of experiments that demonstrated coupling of modeling components from different institutions. ESMF was also used as the basis for the construction of a new model, the Goddard Earth Observing System Model, Version 5 (GEOS-5) atmospheric general circulation model at NASA Goddard.

As the end of the first funding cycle neared, collaborators developed a Project Plan that defined an expanded ESMF organization with multi-agency governance and sponsorship. New five-year grants came from NASA, through the Modeling Analysis and Prediction (MAP) program for Climate Variability and Change, and from the Department of Defense Battlespace Environments Institute. The National Science Foundation (NSF) funded part of the development team through NCAR core funds, and the National Oceanic and Atmospheric Administration (NOAA) provided funding to support operational community requirements. Many smaller ESMF-based application adoption projects were funded in domains as diverse as space weather and sediment transport. During this second project phase, the central data structures in ESMF were completely rewritten to improve flexibility and extensibility.

In 2008, ESMF was chosen as the technical basis for the National Unified Operational Prediction Capability (NUOPC), a consortium of U.S. operational weather and climate centers that aims to deliver an ESMF-based, managed, multi-model ensemble in the 2015 time frame. The emergence of this large-scale national project marks the beginning of the framework's third phase.

5.3 Partnerships and Organization

ESMF governance must be inclusive and responsive to address the needs of its diverse users and sponsors. Core project values include openness of information and development processes.

The ESMF organization is comprised of a Working Project and an Executive Management. The Working Project is defined as the team of customers and developers who collaborate day-to-day to build the ESMF product. The Working Project consists of three parts:

- A line-managed Core Team responsible for building the ESMF software, including unit and system testing, maintenance, support, and oversight of a web-based collaboration environment;
- A group of active users called the Joint Specification Team (JST) that interacts frequently with the Core Team and broader community, providing requirements and feedback;
- A Change Review Board that integrates and prioritizes the requirements from multiple users and sponsors.

The Working Project is funded, guided, and evaluated by its Executive Management which is comprised of several bodies. These are:

- An Executive Board charged with scientific and technical leadership;
- An Advisory Board that reports to and guides the Executive Board;
- An Interagency Working Group that coordinates among sponsors.

Both the Working Project and Executive Management interact with the Earth system modeling and related communities, including the computer science community, the software engineering community, standards bodies, and vendors.

This organization enables the ESMF project to coordinate across milestones, missions, and agencies. Coordination across the ESMF must be addressed by the entire project structure, so that sponsors can coordinate with other sponsors, well-informed users prioritize development tasks, and technical staff and users can *communicate* with their peers.

5.4 Technical Strategy

ESMF is based on principles of component-based software engineering. The components within an ESMF software application usually represent large-scale physical domains such as the atmosphere, ocean, cryosphere, or land surface. Some models also represent specific processes (e.g. ocean biogeochemistry, the impact of solar radiation on the atmosphere) as components. In ESMF, components can create and drive other components so that an ocean biogeochemistry component can be part of a larger ocean component.

ESMF offers two kinds of components: a Gridded Component (GridComp), which is associated with a physical domain, and a Coupler Component (CplComp), for transforming and transferring data between GridComps. ESMF components exchange information with other components only through a State object. A State contains data types representing fields, arrays, or other States. Each Gridded Component is associated with an import State, containing the data required for it to run, and an export State, containing the data it produces.

In order to adopt ESMF, modellers must decide on how to organize their code as a set of GridComps and CplComps, then split these components into standard ESMF methods (initialise, run, and finalize, each of which may have multiple phases). The next step is to wrap native model data structures with ESMF data structures. This can be done either in index space, using a very general ESMF Array class, or in physical space, in which case model grids must be expressed using the ESMF Grid class. If Grids are used, ESMF can generate the interpolation weights needed for regridding between components.

ESMF enables components to run sequentially, concurrently, or in a mixed mode. Applications usually run with all components linked into a single executable program, but there is also support for running separate components as multiple executables. ESMF is written mainly in C++, and has Fortran and C interface bindings.

5.5 Overview of Results

Since its inception in 2002, the ESMF effort has steadily grown, attracting new users, new offshoots, and new sponsors. Its success can be measured by the increasingly robust and fully-featured software that it has delivered, by the growing pool of ESMF components and applications in the community, and by the emergence of new partnerships facilitated by this shared infrastructure.

The most recent release of ESMF, v3.1.1, offers high-level representation and manipulation of observational data streams, unstructured meshes, and logically rectangular grids. Bilinear and higher order interpolation methods are available. The software is supported on more than 30 platform/compiler combinations, and has Fortran and C language bindings.

There are currently about 70 ESMF components and applications in the community. The three largest ESMF systems are the GEOS-5 model at NASA Goddard Space Flight Center, which is structured as a deeply nested component hierarchy; the whole Earth system developed by the Battlespace Environments Institute, which combines coastal, watershed, ocean, atmosphere, and space weather components into multiple models; and the new numerical weather prediction system at the National Centers for Environmental Prediction, which will be a key part of a next-generation operational multi-model ensemble. These activities have been deeply integrative, bringing to bear the resources of multiple organizations on problems too large for any one of them to address alone.

5.6 Future Plans

In the future, ESMF will continue to improve and extend its functionality, improve training materials, and expand and support its customer base.

The project is also evolving to address new concerns. ESMF initially focused on coupling components intended to run on the same computer, with performance as the foremost concern. In response to changing science requirements and technical trends, future plans focus on leveraging the interface and metadata standardisation implicit in ESMF adoption in order to enable ESMF components to operate in more heterogeneous environments. One aspect of this is linking ESMF components to web-based coupling technologies. Another is introducing ESMF components and models into "science gateways" that catalog and integrate diverse, distributed resources.

The integration of modeling with data services is a key part of this vision. The NSF-funded Earth System Curator project, funded in 2005, paired ESMF leads from NCAR and the NOAA Geophysical Fluid Dynamics Laboratory with leads from the Earth System Grid (ESG) data distribution portal, the Georgia Institute of Technology, and the Massachusetts Institute of Technology. Working with the U.K. Numerical Model Metadata (NMM) project, the Curator group is creating a prototype portal based on the ESG and NMM schematas that links information about models, model components, specific simulations, and output datasets. The portal will be used for model intercomparison projects in the climate domain. The extension of this work to support the next Intergovernmental Panel on Climate Change assessment, in coordination with the European Union METAFOR project, is the current focus of follow-on activity for the ESMF team.

Chapter 6
The Grid ENabled Integrated Earth System Modelling (GENIE) Framework

Tim Lenton

The Grid ENabled Integrated Earth system modelling (GENIE) framework is designed: (i) to flexibly couple models of Earth system components (including; ocean, atmosphere, land surface, sea-ice, marine biogeochemistry, terrestrial biogeochemistry, ice sheets), such that they can be readily interchanged at compilation time, (ii) to tune and execute the resulting Earth system models on a wide variety of platforms including across the Grid, and (iii) to archive, query, and retrieve the results.

To date, the emphasis of GENIE has been on computationally efficient models that allow users: (a) to study the full time span of the ice core paleo-climate record (approaching 1 Myr), (b) to make long term future projections, and (c) to conduct extensive searches of model parameter space, undertake comprehensive sensitivity studies, and make probabilistic projections.

The scientific component models (the modules) currently include: (1) Ocean: the Global Ocean Linear Drag Salt & Temperature Equation Integrator (GOLDSTEIN), which is a 3-D frictional geostrophic model with linear drag, (2) Atmosphere: either (a) a 2-D single-layer Energy and Moisture Balance Model (EMBM), or (b) the Reading Intermediate General Circulation Model (IGCM3.1) which is a 3-D spectral primitive equation model, (3) Sea-ice: either (a) dynamic (free-drift) or (b) thermo-dynamic slab, (4) Land surface and carbon cycle: either (a) the Efficient Numerical Terrestrial Scheme (ENTS), (b) the IGCM land scheme, or (c) MOSES-TRIFFID, (5) Marine biogeochemistry: BIOGEM, (6) Marine sediments: SEDGEM, (7) Rock weathering: ROKGEM, (8) Ice sheets: GLIMMER. See Lenton et al. (2007) and references therein for more details of most of these.

T. Lenton
College of Life and Environmental Sciences, University of Exeter, Exeter, UK
e-mail: T.M.Lenton@exeter.ac.uk

K. Puri et al., *Earth System Modelling – Volume 1*, SpringerBriefs in Earth System Sciences, DOI: 10.1007/978-3-642-36597-3_6, © The Author(s) 2013

6.1 Rationale and History

Building the scientific components of an Earth system model requires the work of many people, and the same is true of the overarching software framework. The success (or otherwise) of GENIE relies on a tight-knit core community where the contributors depend on one another. The scientific group began to form around the year 2000 out of the NERC Earth System Modelling Initiative (NESMI), thanks to the leadership of John G. Shepherd. Following some unsuccessful attempts to get funding, the UK e-Science programme provided an ideal opportunity. The scientific group soon realised the benefits of working with computational scientists to achieve our aims to produce flexible, modular, scaleable models. Simon J. Cox and Stephen Newhouse showed how generic software could be used for state-of-the-art model tuning, and how the Grid offered the opportunity for high throughput of simulations through the efficient use of shared resources. Thus, the first phase of the GENIE project (2003–2006) was born, under the leadership of Paul J. Valdes. The initial work focused on the construction and coupling of Earth system model components, as well as experimenting with the potential of the Grid for large ensemble simulations, e.g. Marsh et al. (2004). The second phase of the project (2006–2008), known as GENIE*fy* (developing a Grid ENabled Integrated Earth system modelling *f*ramework for the community) and led by Timothy M. Lenton, has been more outward looking in linking our software framework to others, including PRISM, the Tyndall Centre Community Integrated Assessment System (CIAS, see Chap. 8 in Vol. 5 of this series), and developing the Bespoke Framework Generator (BFG2, see Chap. 7 in Vol. 3 of this series).

6.2 Partnership and Organization

The partnership involves members of the Universities of East Anglia (School of Environmental Sciences), Bristol (Department of Geography), Southampton (Schools of Ocean and Earth Sciences, Engineering Sciences), Reading (Department of Meteorology), Manchester (Centre for Novel Computing), the Open University (Earth Sciences), and Imperial College (Computing). It has also involved members of institutes including the NERC Centre for Ecology and Hydrology (Edinburgh and Wallingford), the National Oceanography Centre (Southampton), and the Hadley Centre for Climate Prediction and Research. There are (or have been) international project partners at the Frontier Research Centre for Global Change (Japan), University of Bern (Switzerland), and University of British Columbia, Vancouver (Canada). With such a distributed group, organization has been aided by quarterly project meetings, including regular use of Access Grid facilities, an active email list, website,[1] and a wiki.[2]

---

[1] http://www.genie.ac.uk

[2] http://source.ggy.bris.ac.uk/wiki/GENIE

6.3 Overview of Results

6.3.1 Version Control, Build and Test

Several software development tools and approaches have been key to making the overall framework development effort coordinated and efficient. A single, central source code repository is used, maintained using a version control system — the open-source Subversion (SVN). In line with good software engineering practice, a policy of frequently testing the source code during development has been adopted. Tests are performed for water conservation, restarts, and the integrity of model runs relative to a suite of standard simulations. To make testing as easy as possible, an integrated build and test system has been created that is stored alongside the source code in the repository. This is particularly challenging for models which embody sensitivity to initial conditions and are developed by researchers working with different compilers and hardware architectures. It is worthwhile because it facilitates rigorous quality control.

6.3.2 Model Coupling and Configuration

At present the default approach to coupling component models uses a master Fortran routine (genie.F) with the model configuration options specified in Extensible Markup Language (XML) files. Two additional software approaches to flexible model coupling have also been developed or experimented with; BFG2 and PRISM. Achieving full modularity poses considerable scientific and technical challenges — for example in reconciling the different time-stepping and flux-exchanges of atmosphere, ocean and sea-ice routines that are present in the parent models that are being modularised. This represents ongoing work and the list of possible combinations is growing. The greatest number of combinations are available in the default (genie.F) approach, whereas only a limited number of configurations have thus far been coupled using BFG2 or PRISM approaches.

genie.F

The master Fortran routine explicitly specifies (i) the order in which component models are called; (ii) the timestepping and relative frequency of the calls; (iii) calls to any required interpolation routines; (iv) a sequential (rather than parallel) execution of model routines; and (v) the respective grid resolutions (through preprocessor statements). BFG2 and PRISM can be used to introduce more flexibility and standardization in (i) through to (iv). Steps to simplify (v) are also underway through the use of run-time, rather than compile-time options by the component models.

BFG2

The Bespoke Framework Generator (BFG) software is described in more detail in Volume 3 of this series. BFG2 was developed to demonstrate a more flexible approach to creating configurations of existing GENIE models and to ease the integration of new models and permit experimentation with alternative model orderings and connectivity. A further aim was to allow the execution of GENIE components either in the form of a single executable running on a single processor, as originally supported by the GENIE framework, or in a distributed computing environment. BFG2 supports a flexible allocation of models to executables and the selection of one of a number of technologies to exchange coupling data between models, for example, 'shared memory' or MPI or the PRISM coupler, OASIS4 (see below).

The BFG2 approach replaces the original GENIE main code, in which run-time switches are used to select a configuration of models and an inflexible model execution order is prescribed. BFG2 generates a bespoke main code from a metadata-level description of a specified model configuration and model ordering. Changing metadata and re-running BFG2 is much more efficient and less error prone than re-writing code directly.

BFG2 provides flexibility in the composition and deployment of coupled models, without requiring modification to the GENIE component models themselves (where the models are viewed as being encapsulated in FORTRAN subroutines). It does this without loss of performance or the introduction of changes to results when compared with the same configuration under the original GENIE framework (i.e. when executed as a single executable; Ford et al. 2007; Armstrong et al. 2009).

Finally, BFG has supported the integration of GENIE with the Community Integrated Assessment System (CIAS). CIAS was originally developed to use BFG1 (Warren et al. 2008) which supported only a direct, point-to-point, description of couplings between models. In the GENIE project, BFG was further extended to support the coupling of models such as those in CIAS, described in the style of BFG1, with GENIE models described in the more flexible style of BFG2.

PRISM

PRISM has been described in Chap. 4. As part of the GENIEfy project, the PRISM coupler OASIS4 (Redler et al. 2010) was used to carry out interpolation between atmosphere, ocean and sea-ice grids in place of the default GENIE interpolation routine. This was achieved by modifying BFG2 so that it could generate OASIS4 communications code and input files. Functionality to read GENIE grid metadata from XML files was also added to BFG2. OASIS4 bilinear, bicubic and nearest neighbour interpolation algorithms were tested, and the flexibility of BFG2 was used to generate multi-executable and single-executable deployments of GENIE. As expected, there was a time overhead to using an external coupler, although little difference was found between interpolation algorithms. There are significant benefits, however, to using OASIS4 combined with BFG2: well-defined regridding methods, including conservative regridding; deployment flexibility allowing GENIE to be run in different ways (including deployments where component models run in parallel); and the ability to add new OASIS4-compatible component models to GENIE in future.

Aladdin

Configuring and using GENIE has been considerably simplified with the development of a graphical user interface (GUI) based on Matlab. This front-end, called 'Aladdin', provides the user with a single window application to access currently available modules and the various compile time and run-time model configurations. Aladdin is designed to aid the user in setting up large ensembles for model tuning and/or sensitivity studies. It also implements some basic error-checking and automatic correction routines.

6.3.3 Grid-Enabled Problem Solving Environment

The collaborative Grid-enabled Problem Solving Environment (PSE) used for composing model studies, accessing distributed computing resources, archiving, sharing and visualizing the results is built upon products of the first phase of the UK e-Science core programme (Hey and Trefethen 2002), in particular the Geodise project.[3] These have been augmented by a set of generic toolboxes for the Matlab and Jython environments (Eres et al. 2005), originally developed to provide solutions for design search and optimisation in aerospace engineering.

Compute Toolbox

The Geodise Compute Toolbox provides intuitive high-level functions in the style of the hosting problem solving environment to allow users to easily manage the execution of a compute job on Grid resource. Functions are provided for three key activities:

- Authentication: In the UK e-Science community users are issued with a X.509 certificate by a trusted Certificate Authority. The toolbox enables the user with such a certificate to create a further time limited proxy certificate which effectively provides a single sign-on to the UK Grid. All subsequent activity on Grid resource is authorised based upon the local rights of the account belonging to the certificate owner. Functionality is provided to instantiate, query and destroy proxy certificates;
- File transfer: The GridFTP (Allcock et al. 2002) data movement service of the Globus Toolkit (2.4) (Foster and Kesselman 1999) is exposed to the Matlab client

---

[3] http://www.geodise.org

as a set of high-level functions. Methods for transferring files to and from a GridFTP enabled resource are provided;

- Job submission: The user is enabled to execute work on resource managed by either the Globus Toolkit or Condor (Thain et al. 2005). By providing information describing the compute task (executable, input files, environment variables) the interface allows submission of jobs to the resource broker of the remote resource. Functions are provided to monitor the status of the job handles returned after submission and to kill those jobs if necessary.

Database Toolbox

The Geodise data management model allows local data (files, scripts, binaries, workspace variables, logical data aggregations) to be archived in a shared central repository and for rich descriptive metadata to be associated with that data. The data can be archived to, queried in, and retrieved from the repository using functions in the user's problem solving environment (PSE). The interface to the database is exposed using Web Services allowing users access to the repository from distributed locations using standard web protocols. Files are stored on a GridFTP server hosted by the UK National Grid Service.[4] The Geodise Database Server has been augmented for GENIE data by mapping an XML Schema into the database to restrict the permissible metadata describing entities in the database. This significantly improves query and retrieval performance in the underlying database. Both programmatic and GUI interfaces are provided to the data repository allowing easy navigation of the data and enabling the database to be an integral part of scripted workflows.

XML Toolbox

A key enabling technology for the management of models from the GENIE framework is the Geodise XML toolbox. This provides a set of functions in the Matlab and Jython environments which convert all common data types that appear in the respective PSEs into a schema-guided XML text structure. It is a very popular toolbox for general data conversion, not just for the specific GENIE application. The toolbox enables users to load the GENIE XML configuration files into the PSE, manipulate the model data (parameters, composition, boundary conditions, etc.) and parse the modified configuration back to XML for further processing. Each XML document output from the PSE in this process completely describes a single simulation and can be archived as metadata in the database and associated with data files generated by the execution of the model. The database's query interface provides the means to interrogate the XML metadata in order to subsequently locate and retrieve data from a particular simulation.

---

[4] http://www.ngs.ac.uk/

OptionsMatlab

The GENIE client for Matlab also includes an interface to a third party design search and optimisation package, OPTIONS (Keane 2003), that has been developed in the Computational Engineering and Design Centre at the University of Southampton. This software provides a suite of sophisticated multidimensional optimisation algorithms developed primarily for engineering design optimisation. The package has been made available to Matlab via the OptionsMatlab interface and has been exploited in conjunction with the Geodise Toolboxes to tune model parameters (Price et al. 2007).

GENIE Toolbox

A higher level abstraction of the Geodise functionality has been developed to provide intuitive management of time-stepping codes on the Grid. Scripted workflows wrapping the Geodise functions have been written to provide a uniform interface for the execution of GENIE Earth system models on local and distributed (Grid) computing resources, the latter using Globus (Foster and Kesselman 1999) and/or Condor (Thain et al. 2005) software. The configuration and execution of a simulation is enabled through a single function call which accepts as input data structures describing the model instance (parameter settings, input files, etc.), the local runtime environment and the remote resource on which to execute the model. Further functionality is provided to coordinate the execution of ensemble studies mediated by the database. The toolbox methods enable users to expose models as tunable functions or include the database as an integral part of a large ensemble study.

Workflow Methods

With the GENIE Toolbox users are able to carry out studies by orchestrating a set of activities using a scripting approach. The project has developed a number of re-usable scripted workflows on tasks including parametric sweeps, parameter estimation and contribution to large collaborative ensemble studies. Scripted workflows enacted in environments such as Matlab and Jython are satisfactory but there are limitations to the nature of the workflows that can be expressed and the robustness with which they can be executed. Microsoft Windows Workflow Foundation has been applied to the GENIE framework to provide a complementary simulation management system capable of reliably managing long-running ensemble simulations (Fairman et al. 2009).

6.4 Users

The UK user base is now approaching 100 people, and there are around 20 international users. They range from PhD students to professors. An educational version of GENIE has been developed by the Open University. GENIE is used in teaching at a number of the partner UK universities.

6.5 Future Plans

With the end of the UK and NERC e-Science programmes efforts are currently under way to seek funding to maintain the existing software and support the ongoing developments being made by the active user base. How the software will be maintained in the long-term remains an open question. Work is ongoing to produce an open source version of most of the GENIE software, with the support of the Open Middleware Infrastructure Institute (OMII). This is beginning with the creation of an improved, open source version of the Aladdin GUI so that GENIE can be freely used for educational purposes.

References

Lenton TM, Marsh R, Price AR, Lunt DJ, Aksenov Y, Annan JD, Cooper-Chadwick T, Cox SJ, Edwards NR, Goswami S, Hargreaves JC, Harris P, Jiao Z, Livina VN, Payne AJ, Rutt IC, Shepherd JG, Valdes PJ, Williams G, Williamson MS (2007) Effects of atmospheric dynamics and ocean resolution on bi-stability of the thermohaline circulation examined using the Grid ENabled Integrated Earth system modelling (GENIE) framework. Clim Dyn 29:591–613

Marsh R, Yool A, Lenton TM, Gulamali MY, Edwards NR, Shepherd JG, Krznaric M, Newhouse S, Cox SJ (2004) Bistability of the thermohaline circulation identified through comprehensive 2-parameter sweeps of an efficient climate model. Clim Dyn 23:761–777

Ford RW, Riley GD, Armstrong CW (2007) Efficient coupling of iterative models. In: Proc. Twelfth ECMWF workshop on the use of high performance computing in meteorology; (George Modzynski, ed.), World Scientific, Reading, England, October 30th-November 3rd 2006, pp 178–190.

Armstrong CW, Ford RW, Riley GD (2008) Coupling integrated Earth System Model components with BFG2. Concurrency Comput: Pract experience 21(6):767–791. doi:10.1002/cpe.1348

Warren R, de la Nava Santos S, Arnell NW, Bane M, Barker T, Barton C, Ford R, Füssel HM, Hankin RKS, Klein R, Linstead C, Kohler J, Mitchell TD, Osborn TJ, Pan H, Raper SCB, Riley G, Schellnhuber HJ, Winne S, Anderson D (2008) Development and illustrative outputs of the Community Integrated Assessment System (CIAS), a multi-institutional modular integrated assessment approach for modelling climate change. Environ model softw 23(5):592–610. doi:10.1016/j.envsoft.2007.09.002

Redler R, Valcke S, Ritzdorf H (2010) OASIS4 - A coupling software for next generation earth system modelling. Geosci Model Dev 3:87–104. doi:10.5194/gmd-3-87-2010

Hey T, Trefethen AE (2002) The UK e-science core programme and the grid. Future Gener Comput Syst 18(8):1017–1031. doi:10.1016/S0167-739X(02)00082-1

Eres MH, Pound GE, Jiao Z, Wason JL, Xu F, Keane AJ, Cox SJ (2005) Implementation and utilisation of a grid-enabled problem solving environment in matlab. Future Gener Comput Syst 21(6):920–929. doi:10.1016/j.future.2003.12.016

Allcock B, Bester J, Bresnahan J, Chervenak AL, Foster I, Kesselman C, Meder S, Nefedova V, Quesnal D, Tuecke S (2002) Data management and transfer in high performance computational grid environments. Parallel Comput J 28(5):749–771

Foster I, Kesselman C (1999) Globus: a toolkit-based grid architecture. In: Foster I, Kesselman C (eds) The grid: blueprint for a new computing infrastructure. Morgan Kaufman Publishers, Inc., San Francisco, USA, pp 259–278

Thain D, Tannenbaum T, Livny M (2005) Distributed computing in practice: the condor experience. Concurrency and Comput: Pract Experience 17(2–4):323–356. doi:10.1002/cpe.938

Keane AJ (2003) The OPTIONS design exploration system: reference manual and user guide. http://www.soton.ac.uk/ajk/options.ps

Price AR, Xue G, Yool A, Lunt DJ, Valdes PJ, Lenton TM, Wason JL, Pound GE, Cox SJ (2007) Optimisation of integrated earth system model components using grid-enabled data management and computation. Concurrency and Comput: Pract and Experience 19(2):153–165. doi:10.1002/cpe.1046

Fairman MJ, Price AR, Xue G, Molinari M, Nicole DA, Marsh R, Lenton TM, Takeda K, Cox SJ (2008) Earth system modelling with windows workflow foundation. Future Gener Comput Syst 25(5):586–597. doi:10.1016/j.future.2008.06.011

Chapter 7
K-1(Kyosei) Project

Akimasa Sumi

7.1 General Overview

High-resolution coupled atmosphere-ocean models are necessary in order to obtain reliable information about the influence and impact of global warming. A high-end computing infrastructure is indispensable to run such models and from 1998 to 2002 the Earth Simulator with a computational speed of 40 TFLOPS was developed in Japan. Subsequently, a model development project known as "Kyosei", which means "co-existence" between human beings and the Earth environment in Japanese, was started in 2002 to develop models suitable to run on the Earth Simulator. The aim of the first Kyosei project (denoted as K-1) was to develop a high-resolution coupled atmosphere ocean model, and included researchers from the Center of Climate System Research (CCSR), National Institute for Environmental Studies (NIES) and Japan Agency for Marine-Earth Science and Technology (JAMSTEC). A brief summary of the project is provided in the following sections.

7.2 Rationale and History

Following wide-ranging discussions in 1998 concerning future supercomputing architectures, a vector-parallel architecture with a large grain was selected as it had advantages for handling fluid dynamical equations. Thus, the Earth Simulator was developed and was used for weather and climate simulations. It consisted of 640 nodes, which utilize 8 processors per node. Each processor speed was 8GFLOPS, giving a total performance of 40 TFLOPS (640 × 8 × 8). A model development

A. Sumi
Integrated Research System for Sustainability Science, Transdisciplinary Initiative for Global Sustainability (TIGS), Atmosphere and Ocean Research Institute (AORI), The University of Tokyo, Tokyo, Japan
e-mail: sumi@ccsr.u-tokyo.ac.jp

K. Puri et al., *Earth System Modelling – Volume 1*, SpringerBriefs in Earth System Sciences, DOI: 10.1007/978-3-642-36597-3_7, © The Author(s) 2013

project was started at the same time to create an advanced climate model suitable
for this newly developed computing architecture. A two-track approach was adopted
consisting of fast and slow tracks. In the fast track, a climate model was developed as
an extension of the exisiting architecture and comprised of (1) a hydrostatic approximation, (2) a global spectral method, and (3) a grid-point ocean model. In this track,
the results were expected to be obtained within 1–2 years of the completion of the
Earth simulator. This initial development was followed by the development of a high
resolution atmosphere-ocean coupled model. The model is known as the Model for
Interdisciplinary Research On Climate (MIROC, Hasumi and Emori 2004), and the
results from the model have appeared in IPCC AR4 and scientific journals (Solomon
et al 2007; Sumi 2008). In the slow track approach, the aim was to develop a next-
generation global model. Key features of the new model included (1) non-hydrostatic
approximation and (2) icosahedral grid atmospheric model. This model is known as
the Nonhydrostatic ICosahedral Atmospheric Model (NICAM, Satoh et al (2008)
). NICAM has now become a "cloud-permitting atmospheric general circulation
model" and is utilized for various research topics. For example, Madden-Julian oscillation was successfully simulated (Miura et al 2007) and a tendency of typhoon
activity in the global warming was investigated by Yamada et al (2010).

7.3 Partnership and Organization

The model development under the Kyosei programme encompassed seven projects.
The first Kyosei project (denoted as K-1) was to develop a high-resolution coupled
atmosphere-ocean model. The second and fourth projects were aimed at developing a carbon-cycle model and a 20 km Atmospheric General Circulation Model
respectively. Other projects were for carbon cycle, water-cycle water-resource management and data assimilation. The K-1 project involved researchers from CCSR,
NIES and JAMSTEC. The K-2 project was coordinated by JAMSTEC while the K-4
project was coordinated by the Meteorological Research Institute (MRI) of the Japan
Meteorological Agency (JMA).

7.4 Overview of Results

7.4.1 Programming Structure

Model development was conducted in order to use the Earth Simulator efficiently.
This required particular attention to both parallel and vector efficiency. The climate
model included a spectral atmospheric general circulation model (AGCM) and a grid
point ocean general circulation model (OGCM). A Multiple Program and Multiple
Data (MPMD) approach was used for coupling as the AGCM and OGCM used different horizontal resolutions. Thus different numbers of nodes were allocated to the

AGCM and OGCM, to be run independently in each node. In order to minimise waiting times for the exchange of data between the AGCM and OGCM, computational load had to be balanced on each node. As a result 10 nodes were allocated to the AGCM and 80 nodes (later, reduced to 76 nodes to save CPU time) to the OGCM. All data had to be gathered on one node as the coupling between the AGCM and the OGCM was conducted on one node. This process takes more computational time than coupling on each node and will be addressed in the next version.

Domain decomposition is one of the key issues in parallel computing and a one-dimensional decomposition was used in the AGCM with a 2-dimensional decomposition in the OGCM. For transformation between grid-space and wave-space in the AGCM, arrays had to be rearranged, to allow Fast Fourier Transforms to be used in the zonal direction and Legendre transformation in the meridional direction.

7.4.2 Scientific Results

The modelling system has delivered a number of interesting scientific results and many of these were included in the IPCC AR4. The general features of the high-resolution climate model are summarized by Sumi (2008) and some interesting results are summarized such as:

- A strong interaction between Hawaiian island mountains, atmosphere and the upper ocean was noted (Sakamoto et al 2004). A similar interaction is noted in the Caribbean Sea and other coastal areas;
- The Kuroshio current around Japan was well reproduced for the first time (see Fig.1 or Fig. 8.1; Solomon et al 2007; Sakamoto et al 2005);
- 20th century climate change was well reproduced and the contribution of human activity is clearly shown (Nozawa et al 2005).

7.5 Users

Since its establishment by the University of Tokyo in 1991, CCSR has been the center for atmospheric and oceanic modelling in the Japanese university community. It has educated many Ph.D students, who have an interest in developing and using advanced models of the atmosphere, ocean and climate.

The AGCM and OGCM developed at CCSR have been widely used in the university community in Japan. Besides domestic collaboration, international collaboration was also pursued. For example, the radiation code in the CCSR/AGCM was exported to Seoul National University (SNU), Korea in the early 90s, and it was used in the SNU AGCM. Both CCSR and MIROC models participated in the IPCC AR4. The MIROC data, which is included in the Coupled Model Intercomparison Project 3 (CMIP3) data archive, is being used by many researchers around the world.

7.6 Future Plans

The follow-up project to Kyosei, known as the "Kakushin" programme, commenced in 2008 following the successful completion of the former project. In this project model development has been continued with an emphasis on greater integration with other disciplines. Thus, for example, integration of model development and application to society is emphasized. Two target timescales have been introduced; (1) long-term simulations, where simulations over 100–300 year are targeted and a carbon-cycle will be developed, and (2) near-term prediction, where future predictions out to 20–30 years are conducted. The K-1 project has been continued in the Kakusin-2. The strategy of a family of models will be used involving a combination of a high-resolution model and a medium-resolution model with the same physical packages. Research results will be presented in IPCC AR5.

The current high resolution coupled model will be improved as follows:

- Horizontal resolution will be increased from T106 to T213 in the atmospheric component;
- The same grid size (approximately 20km) will be used in the ocean component but a tri-polar grid will be introduced;
- New physical processes will be implemented.

The current medium resolution model will also be improved as follows:

- Horizontal resolution will be increased from T42 to T85 in the AGCM;
- The same ocean component will be used;
- Physical processes will be the same as in the high-resolution model.

Besides this coupled model, a cloud-permitting NICAM (7 km or 3.5 km) will be developed for seasonal prediction and down-scaling purposes.

References

Hasumi H, Emori S (2004) K-1 Coupled Model(MIROC) description. Technical Report K-1, CCSR, University of Tokyo.

Solomon S, Qin D, Manning M, Chen Z, Marquis M, Averyt KB, Tignor M, Miller HL (2007) Contribution of working group I to the fourth assessment report of the intergovernmental panel on climate change. Technical report, Cambridge, United Kingdom and New York, NY, USA, http://www.ipcc.ch/publications_and_data/ar4/wg1/en/contents.%html

Sumi A (2008) Global warming simulation using the high-resolution climate model: a summary of the K-1 project. J Earth Simulator 9:37–49

Satoh M, Matsuno T, Tomita H, Miura H, Nasuno T, Iga S (2008) Nonhydrostatic icosahedral atmospheric model (NICAM) for global cloud resolving simulations. J Comput Phys 227(7): 3486–3514, doi:10.1016/j.jcp.2007.02.006

Miura H, Satoh M, Nasuno T, Noda AT, Oouchi K (2007) A madden-julian oscillation event realistically simulated by a global cloud-resolving model. Science 318(5857):1763–1765, doi:10.1126/science.1148443

Yamada Y, Oouchi K, Satoh M, Tomita H, Yanase W (2010) Projection of changes in tropical cyclone activity and cloud height due to greenhouse warming: global cloud-system- resolving approach. Geophys Res Lett 37:L07,709. doi:10.1029/2010GL042518

Sakamoto TT, Sumi A, Emori S, Nakajim T, Suzuki T, Kimoto M (2004) Far-reaching effects of the Hawaiian Islands in the CCSR/NIES/FRCGC high-resolution climate model. Geophys Res Lett 31:L17,212

Sakamoto TT, Hasumi H, Ishii M, Emori S, Suzuki T, Nishimura T, Sumi A (2005) Responses of the Kuroshio and the Kuroshio extension to global warming in a high-resolution climate model. Geophys Res Lett 32:L14,617. doi:10.1029/2005GL023384

Nozawa T, Nagashima T, Shiogama H, Crook SA (2005) Detecting natural influence on surface air temperature change in the early twentieth century. Geophys Res Lett 32:L20,719

Chapter 8
CCSM

Mariana Vertenstein, Anthony Craig and Robert Jacob

8.1 General Overview

The Community Climate System Model (CCSM) development and coordination is based at the National Center for Atmospheric Research (NCAR) in Boulder, Colorado, USA. CCSM is a state-of-the-art global climate model consisting of four fundamental physical components: an atmosphere model, an ocean model, a land surface model, and a sea ice model. In addition, a coupler (or driver) is used to exchange boundary data between the components and to coordinate the time evolution of the physical models. CCSM is used to understand the Earth's global climate system, to predict the effects of climate change, and to understand past climates. It is developed as a high performance computing application but is used on a wide variety of platforms.

8.2 Rationale and History

CCSM development began in the mid 1990s as the NCAR Climate System Model (CSM). In 2000, community was added to the project name even though the community was involved in development in the first few years. In 2010, the Community Climate System Model became the Community Earth System Model with

M. Vertenstein · A. Craig
National Center for Atmospheric Research, Boulder, CO, USA
e-mail: mvertens@ucar.edu

A. Craig
e-mail: tcraig@ucar.edu

R. Jacob
Argonne National Laboratory, Chicago, IL, USA
e-mail: jacob@mcs.anl.gov

K. Puri et al., *Earth System Modelling – Volume 1*, SpringerBriefs in Earth System Sciences, DOI: 10.1007/978-3-642-36597-3_8, © The Author(s) 2013

the introduction of many new biogeochemistry capabilities and other features. CCSM benefits from a long history of NCAR community-type model development activities such as CCM (Hack et al. 1998), a coupled atmosphere model; PCM (Washington et al. 2000), a massively parallel coupled climate model; and LSM (Bonan 1998), a community land surface model.

Several versions of CCSM have been released over the history of the project. Three versions of CCSM1 (Boville and Gent 1998) were released in 1996, 1998, and 2000. CCSM2 (Kiehl and Gent 2004) was released in 2002, CCSM3 (Collins et al. 2006a) was released in 2004, and CCSM4 was released in early 2010. The first version of the Community Earth System Model (CESM1) which adds an additional land-ice component and adds biogeochemistry capabilities was released in mid-2010. The versions have evolved rapidly over the last decade with continual improvements in both scientific fidelity as well as technical capability. CCSM1 ran only on high performance CRAY shared memory vector platforms. With CCSM2, the Message Passing Interface (MPI) was formally adopted and the system ran on a greater variety of both high performance distributed and shared memory systems. CCSM3 improved scalability and with the CCSM4 release, CCSM runs on an even greater variety of platforms ranging from single processor laptop machines to "petaflop" machines at 100,000 processors or more.

Like many climate models, CCSM is not developed from scratch as a single application but is instead a coupled application where each model component (atmosphere, ocean, land, and sea ice) is a highly complex application developed separately with its own coding style and datatypes. The CCSM fully active physical components have evolved significantly over the history of the project. In CCSM4, the active components are based on the Community Atmosphere Model (CAM; Collins et al. 2006b), the Community Land Model (CLM; Bonan et al. 2002), the Parallel Ocean Program (POP; Smith and Gent 2004), and the Los Alamos Sea Ice Model (CICE; Hunke and Dukowicz 1997). Each of these models have been developed with significant community involvement, and many components or component features were developed primarily outside NCAR within the community.

The CCSM community has contributed significant scientific, technical, and computational resources from outside NCAR to benefit the project. In return, the CCSM project is able to provide the community with both a flexible and robust publicly available climate model as well as results from a significant numbers of experimental cases for analysis and comparison.

8.3 Partnerships and Organization

The CCSM project is sponsored by the National Science Foundation (NSF) and the Department of Energy (DOE). CCSM is a community model and as a result has partnerships with many collaborators. In particular, over the last several years, many groups have contributed significantly to various aspects of scientific and technical development as well as validation. These include groups from within the University community; the DOE national laboratories including Argonne (ANL), Los Alamos (LANL), Lawrence Berkeley (LBNL), and Oak Ridge (ORNL) National Laboratory; NASA through partnerships with the Earth System Model Framework (ESMF) Project; and others.

CCSM is organized at NCAR as a community project with control of the project managed by a steering committee of about 15 people, half from NCAR and half from outside NCAR. That committee has an overall head that is usually a scientist at NCAR. CCSM also has an external advisory board that reviews progress and plans and makes recommendations. The board consists of approximately 10 people, all external to NCAR, and that board consists of both major sponsers of the project and leaders in the field of climate research that are not active participants in CCSM.

The CCSM project is organized into about 12 working groups in areas such as atmosphere, land, sea ice or ocean model development; climate change; biogeochemistry; paleo-climate; or software engineering. Each group has two coordinators, one from NCAR and one from the outside community, and the groups welcome participation from anyone in the community.

8.4 Overview of Results

Over the past fifteen years CCSM has developed a coupling infrastructure that permits coupling with minimum modification to individual component models and that has largely been kept separate from the basic component model physics implementation. Prior to CCSM4 and as documented in greater detail in Craig et al. (2005), the model operated as a multiple executable system where all models ran concurrently as separate binary executables on non-overlapping processor sets and there was no concept of a top-level driver. The components started independently and sent and received boundary data to and from the coupler at regular intervals via send and receive methods called directly from within each component.

The coupler acted as a central hub coordinating data exchange, implicitly managing lags and sequencing, and executing coupler operations such as mapping (interpolation) and merging. The boundary data exchanges in CCSM3 were implemented in a way that met the scientific requirements of the system while maximising the amount of concurrency between components. This fully concurrent approach had not fundamentally changed since the development of CCSM1 in the mid-1990s.

With CCSM4, a completely new approach has been taken with respect to the high-level architecture and design of the system. The CCSM4 system is a single executable system that provides new flexibility in running model components sequentially, concurrently, or in a mixed sequential/concurrent mode. To achieve this additional flexibility, a new driver has been introduced in the CCSM system. The driver runs on all the processors of a model simulation and controls the time sequencing, processor concurrency, and exchange of boundary data between components. In particular, the driver calls all model components via common and standard interfaces and also directly calls coupler methods for mapping (interpolation), rearranging, merging, an atmosphere/ocean flux calculation, and diagnostics. In some sense, the CCSM3 sequencing and hub attributes have been migrated to the driver level while the coupler operations like mapping and merging are being done as if a separate coupler component existed in CCSM4.

In CCSM4, all model components and the coupler methods can run on subsets of all the processors. This permits the model system to have greatly increased flexibility in setting up an appropriate component layout to optimise the performance of a model simulation. CCSM4 supports an all concurrent layout. It also supports component layouts where all components are running on the same processors sequentially, but CCSM4 more typically runs optimally in a mixed sequential concurrent layout of components where some components are running sequentially and others are running concurrently. In CCSM4, the component processor layouts and MPI communicators are derived from a run time input. The driver derives all MPI communicators at initialisation and passes them to the component models for use. All CCSM4 components are parallelised using MPI and many also support the use of OpenMP as an additional level of parallelism.

As much as possible, the CCSM4 driver sequencing has been implemented to maximise the potential amount of concurrency of work between different components. Ideally, in a single coupling step, the forcing for all models would be computed first, the models could then all run concurrently, and then the driver would advance. However, scientific requirements in CCSM4 prevent this ideal implementation. In order to resolve the diurnal cycle properly, the land, sea ice, and ocean albedo must be computed for the next radiation timestep before the atmosphere model radiation is updated. In addition, the lagging of the model state relative to the albedo computation must be minimised due to stability considerations. In CCSM4, results are bit-for-bit identical regardless of the component processor layout because the scientific lags are fixed by the driver implementation and not the processor layout.

Depending on the resolution, hardware, run length and physics, a CCSM4 run can take several hours to several months of wall time to complete. Runs are typically decades or centuries long, and the model typically runs between 1 and 50 model years per wall clock day. CCSM4 has exact restart capability and the model is typically run in individual 1 year or multi-year chunks. CCSM4 has automatic resubmission and automatic data archiving capability.

The Argonne National Laboratory (ANL) Model Coupling Toolkit (MCT, see Volume 3, Chap. 3 of this series and Larson et al. 2005) was a critical piece of software in the CCSM3 coupler. In the updated CCSM4 driver, the MCT attribute_vector, global_segmap, and general_grid datatypes have been adopted at the highest levels of the driver, and they are used directly in the standard component interfaces. In addition, MCT is used for all data rearranging and mapping (interpolation). The clock used by CCSM4 at the driver level is based on an Earth System Model Framework (ESMF) specification (Hill et al. 2004). Mapping weights are still generated external to the model using the SCRIP (Jones 1998) package. They are read into CCSM4 using a custom built subroutine that reads and distributes the mapping weights in reasonably small chunks to minimise the memory footprint. Development of the CCSM4 driver not only relies on MCT, but colleagues at ANL contributed significantly to the design and implementation. Development of both the CCSM3 and CCSM4 couplers has resulted from a particularly strong and close collaboration between NCAR and ANL.

The current CCSM4 component model interfaces are based upon the ESMF standard. Each component provides an init, run, and finalize method with well defined arguments. The CCSM4 interface arguments currently consist of Fortran and MCT datatypes, but alternative ESMF component compliant interfaces are also available. The coupling fields are passed through the interfaces in the init, run, and finalize phases. The driver acquires all information about resolution, configurations, and processor layout at run-time from either namelist input or communication with components.

CCSM4 has targeted much higher resolutions than any previous CCSM coupled model. Efforts have been made to reduce the memory footprint and improve scaling in all components with a target of being able to run the fully coupled system at one tenth degree global resolution on tens-of-thousands of processors on machines where each processor may have as little as 512 Mb of memory. The memory limitations have imposed new constraints on component model initialisation, and significant refactoring has been required in some models within the CCSM community to improve I/O performance and memory usage in the model. In addition, CCSM is using a new parallel I/O library, PIO, layered on top of NetCDF, pNetCDF, and MPI-IO to further improve I/O performance and memory usage in the model.

8.5 Users

CCSM is fundamentally a community model. Both development and use are community efforts. Collaboration between NCAR and outside institutions is strongly encouraged, and participation by anyone in the community welcome. A CCSM Community Meeting is held each year where scientists, developers, and users meet to discuss model development activities and scientific results. Several hundred collaborators usually attend. An updated version of the CCSM model is publicly released every few years via the web, and it is available to anyone for download. This download includes source code, basic model documentation, and model output from many experimental runs.

8.6 Future Plans

In the future, CCSM in the form of CESM will continue to push forward with scientific improvements in areas like subgrid parameterisations, physics packages, and biogeochemistry modeling. Improving the CCSM scientific accuracy is always high priority. To achieve the goals of improving regional climate forecasts and improving scientific accuracy, CCSM will continue to update the scientific packages, begin implementing nested higher resolution models, and increase resolution. On a technical level, CCSM will have to continue to focus on improved memory and performance scalability as a top priority. Finally, as model resolution and computational capability increase, managing model output will continue to become more critical as data generated by the model increases from terabytes to petabytes.

References

Hack JJ, Kiehl JT, Hurrell JW (1998) The hydrologic and thermodynamic characteristics of the NCAR CCM3. J Clim 11:1179–1206

Washington WM, Weatherly JW, Meehl GA, Semtner AJ Jr, Bettge TW, Craig AP, Strand WG Jr, Arblaster JM, Wayland VB, James R, Zhang Y (2000) Parallel climate model (PCM) control and transient simulations. Clim Dyn 16(10/11):755–774

Bonan GB (1998) The land surface climatology of the NCAR land surface model coupled to the NCAR community climate model. J Clim 11:1307–1326

Boville BA, Gent PR (1998) The NCAR climate system model, version one. J Clim 11: 1115–1130

Kiehl J, Gent PR (2004) The NCAR climate system model, version two. J Clim 17:3666–3682

Collins WD, Bitz CM, Blackmon ML, Bonan GB, Bretherton CS, Carton JA, Chang P, Doney SC, Hack JJ, Henderson TB, Kiehl JT, Large WM, McKenna DS, Santer BD, Smith RD (2006a) The community climate system model version 3 (CCSM3). J Clim 19(11):2122–2143

Collins WD, Rasch PJ, Boville BA, Hack JJ, McCaa JR, Williamson DL, Briegleb BP, Bitz CM, Lin SJ, Zhang M (2006b) The formulation and atmospheric simulation of the community atmosphere model version 3 (CAM3). J Clim 19(11):2144–2161

Bonan GB, Oleson KW, Vertenstein M, Levis S, Zeng X, Dai Y, Dickinson RE, Yang XL (2002) The land surface climatology of the community land model coupled to the NCAR community climate model. J Clim 15:3123–3149

Smith RD, Gent PR (2004) Reference manual for the parallel Ocean program (POP): Ocean component of the community climate system model (CCSM2.0 and 3.0). Technical report LA-UR-02-2484, Los Alamos National Laboratory, Los Alamos, NM, http://climate.lanl.gov/Software/SCRIP

Hunke EC, Dukowicz JK (1997) An elastic-viscous-plastic model for sea ice dynamics. J Phys Oceanogr 27:1849–1867

Craig AP, Jacob R, Kauffman B, Bettge T, Larson J, Ong E, Ding C, He Y (2005) CPL6: the new extensible high performance parallel coupler for the community climate system model. Int J High Perform Comput Appl 19(3):309–328

Larson J, Jacob R, Ong E (2005) The model coupling toolkit: a new Fortran90 toolkit for building multiphysics parallel coupled models. Int J High Perform Comput Appl 19(3): 277–292

Hill C, Deluca C, Balaji V, Suarez M, da Silva A (2004) Architecture of the Earth system modeling framework. Comput Sci Eng 6(1):18–28

Jones PW (1998) A user's guide for SCRIP: a spherical coordinate remapping and interpolation package. Technical report, Los Alamos National Laboratory, Los Alamos, NM, http://climate.lanl.gov/Software/SCRIP

Glossary

ACCESS	Australian Community Climate and Earth System Simulator
AGCM	Atmosphere General Circulation Model
ANL	Argonne National Laboratory
APPCIM	Application schema for CIM
AR	Assessment Report
BADC	British Atmospheric Data Centre
BFG	Bespoke Framework Generator
CAM	Community Atmosphere Model
CCM	Community Climate Model
CCSM	Community Climate System Model
CCSR	Center of Climate System Research
CDO	Climate Data Operator
CERFACS	Centre Européen de Recherche et Formation Avancée en Calcul Scientifique
CESM	Community Earth System Model
CF	Climate and Forecast
CIAS	Community Integrated Assessment System
CICE	Los Alamos Sea Ice Model
CICS	Cooperative Institute for Climate Science
CIM	Common Information Model
CLM	Community Land Model
CMIP3	Coupled Model Intercomparison Project phase 3
CMIWG	Common Modelling Infrastructure Working Group
COMBINE	Comprehensive Modelling of the Earth System for Better Climate Prediction and Projection
CONCIM	Conceptual model of the process for CIM
CORDEX	Coordinated Regional climate Downscaling Experiments
COSMOS	COmmunity earth System MOdelS
CSIRO	Commonwealth Scientific and Industrial Research Organisation
DKRZ	German Climate Computing Centre
DOE	Department of Energy
ECMWF	European Centre for Medium-Range Weather Forecasts
EMBM	Energy and Moisture Balance Model

ENES	European Network for Earth System Modelling
ENTS	Efficient Numerical Terrestrial Scheme
ESG	Earth System Grid
ESMF	Earth System Modeling Framework
ESTO	Earth Science Technology Office
FCM	Flexible Configuration Management
GENIE	Grid ENabled Integrated Earth system modelling
GENIEfy	GENIE framework for the community
GEOS—5	Goddard Earth Observing System Model, Version 5
GFDL	Geophysical Fluid Dynamics Laboratory
GFLOPS	Giga Floating Point Operations per Second
GOLDSTEIN	Global Ocean Linear Drag Salt & Temperature Equation Integrator
GUI	Graphical User Interface
HPC	High Performance Computing
IGCM	Intermediate General Circulation Model
IPCC	Intergovernmental Panel on Climate Change
IPSL	Institut Pierre-Simon Laplace
IS—ENES	Infrastructure for the European Network for Earth System Modelling
JAMSTEC	Japan Agency for Marine-Earth Science and Technology
JMA	Japan Meteorological Agency
LANL	Los Alamos National Laboratory
LBNL	Lawrence Berkeley National Laboratory
LSM	Land Surface Model
MAP	Modeling Analysis and Prediction
MCT	Model Coupling Toolkit
METAFOR	Common Metadata for Climate Modelling Digital Repositories
MIROC	Model for Interdisciplinary Research On Climate
MPI	Message Passing Interface
MPI—ESM	Max Planck Institute—Earth System Model
MPI—M	Max-Planck-Institut für Meteorologie
MPMD	Multiple Program and Multiple Data
MRI	Meteorological Research Institute
NASA	National Aeronautics and Space Administration
NCAR	National Center for Atmospheric Research
NCAS	National Centre for Atmospheric Science
NEMO	Nucleus for European Modelling of the Ocean
NERC	National Environment Research Council
NESMI	NERC Earth System Modelling Initiative
NetCDF	Network Common Data Form
NICAM	Nonhydrostatic ICosahedral Atmospheric Model
NIES	National Institute for Environmental Studies
NMM	Numerical Model Metadata
NOAA	National Oceanic and Atmospheric Administration

NSF	National Science Foundation
NUOPC	National Unified Operational Prediction Capability
NWP	Numerical Weather Prediction
OASIS	Ocean Atmosphere Sea Ice Soil (coupler)
OGCM	Ocean General Circulation Model
ORNL	Oak Ridge National Laboratory
PAE	PRISM Area of Expertise
PALM	Projet d'Assimilation par Logiciel Multiméthodes
PCM	Parallel Climate Model
PCMDI	Program for Climate Model Diagnosis and Intercomparison
PIO	Parallel IO
pNetCDF	parallel NetCDF
POP	Parallel Ocean Program
PRACE	Partnership for Advanced Computing in Europe
PRISM	Partnership for Research Infrastructures in earth System Modelling
PSE	Problem Solving Environment
SME	Small and Medium Enterprises
SMS	Supervisor Monitor Scheduler
SOA	Service Oriented Architecture
SVN	Subversion
TFLOPS	Tera Floating Point Operations per Second
THREDDS	Thematic Realtime Environmental Distributed Data Services
UML	Unified Modeling Language
XML	Extensible Markup Language

Index

K. Puri et al., *Earth System Modelling – Volume 1*, SpringerBriefs in Earth System
Sciences, DOI: 10.1007/978-3-642-36597-3, © The Author(s) 2013